Engineering drawing for technicians volume 1

O. Ostrowsky

B.A., M.S.E., P.Eng., Dip. Eng. Design, M.I.E.D., M.I.Plant E.
Lecturer responsible for engineering drawing and design,
People's College for Engineering and Science,
Nottingham

Edward Arnold

Preface

This book completely covers the objectives of the level-1 Technician Education Council standard units in engineering drawing (U75/003 and U75/041), with the following additional topics:

a) perspective projection, which is essential for learning how to sketch correctly;

b) reading an engineering drawing, which is vital for learning to visualise three-dimensional objects from two-dimensional representations.

There is also an explanatory section on technical terms that are commonly used in general engineering, and the relevant parts of BS 308 and BS 4500A are reproduced by kind permission of the British Standards Institution, 2 Park St, London W1A 2BS, from whom copies of the complete standards may be obtained.

More than 100 test questions are included, with over 200 drawing examples suitable for phase tests, final tests, classwork, or homework. Solutions are given to 36 selected questions, and these include more than 160 individual solutions.

For measuring purposes, some examples incorporate construction squares until orthodox dimensioning is properly explained.

If the book is to be used as a class work book, office paper of 45 g/m² grade will provide an economical substitute for tracing paper.

Finally, for simplicity some drawing-office personnel are described by the usual term 'draughtsmen' rather than as 'draughtsmen and draughtswomen' or 'draughtspersons'.

O. Ostrowsky

Contents

1 Communication 1
Engineering drawings, *1*. Main types of engineering drawing, *1*. Drawing equipment, *2*. Layout of drawings, *3*. Lines, *4*. Lettering and numerals, *5*. Technical terms, *6*. Abbreviations and symbols, *7*. Conventional representation of common features, *8*

2 Organisation 10
Stages in the development of a new product, *10*. Standard parts, 10. The drawing office, *10*. The print room, *10*. Drawing-office personnel, *11*. Test questions, *12*

3 Sketching 14
Freehand sketching, *14*. Form and proportion, *14*

4 Pictorial projection 15
Perspective projection, *15*. Test questions, *16*. Isometric projection, *18*. Test questions, *21*. Oblique projection, *23*. Test questions, *25*

5 Orthographic projection 26
First-angle projection, *26*. Test questions, *28*. Third-angle projection, *31*. Test questions, *32*. Sectional views, *38*. Test questions, *41*. Views on drawings, *45*. Test questions, *47*. Points, lines, and plane surfaces in space, *48*. Test questions, *50*. Auxiliary views, *52*. Test questions, *53*

6 Dimensioning 56
Functional dimensioning. *56*. Non-functional dimensioning. *56*. Auxiliary dimensions, *56*. Principles of dimensioning, *56*. Dimensioning for different purposes, *60*. Dimensioning for primary production, *60*. Dimensioning for secondary production, *60*. Limits and fits, *62*. Machining symbol, *67*. Test questions, *67*

7 Fasteners 77
Screw threads, 77. Temporary fastenings, *78*. Permanent fastenings, *81*. Locking devices, *83*. Test questions, *84*

Appendix: drawing ellipses 87

Selected solutions 88

Index 94

Acknowledgements

I wish to express thanks to my wife Catherine, her sister Eileen, and my daughters Sharon and Lisa for their continuing patience and occasional assistance during the preparation of this book.

I am also indebted to my colleagues who read the manuscript and offered helpful criticism.

O.O.

© O. Ostrowsky 1979

First published 1979
by Edward Arnold (Publishers) Ltd
41 Bedford Square, London WC1B 3DQ

ISBN 0 7131 3408 9

Text set in 10/12 pt IBM Press Roman, printed and bound in Great Britain at The Pitman Press, Bath

1 Communication

Throughout the ages, people have found communication with each other to be essential to their development. The means they have used have progressed from grunts to articulate speech and from signs and primitive drawings to competent writing and complicated drawings. All these have served to convey ideas, information, and instructions from one person to another.

In present-day industry, the principal means of communication is *engineering drawing*, which is the international language of engineering.

Engineering drawing is a system of communication in which ideas are expressed exactly, information is conveyed completely and unambiguously, and even the most complicated shapes are specifically described.

In Great Britain, the international conventions of engineering drawing are published by the British Standards Institution in British Standard BS 308: 1972, 'Engineering drawing practice'. This standard enables the draughtsman to understand clearly the designer's ideas and instructions and the craftsman to interpret precisely an engineering drawing for manufacturing or assembly purposes.

1.1 Engineering drawings

Engineering drawings are two-dimensional visual representations of three-dimensional objects and are used as a universal means of communication in industry.

Such drawings must be clear, concise, and accurate. They should convey, when required,
a) information about the shapes and sizes of components,
b) material requirements, and
c) instructions about the method of manufacture.
All information must be complete and specified once only.

1.2 Main types of engineering drawing

Single-part or component drawings (figs 1.1 to 1.3)
Single-part detail drawings are usually the last in a series of different drawings.

Detail drawings should contain sufficient information for the production department to make the components. This should include
a) dimensions and tolerances,
b) material specifications,
c) manufacturing processes and machining instructions,
d) heat-treatment instructions etc.

Sub-assembly drawings
Sub-assembly drawings show the arrangement of several adjacent parts which together form a part of the finished assembled product. These drawings may include fitting dimensions and a parts list, which usually contains the following information:
a) item numbers to identify the parts comprising the assembly,
b) descriptions, and
c) quantities required.

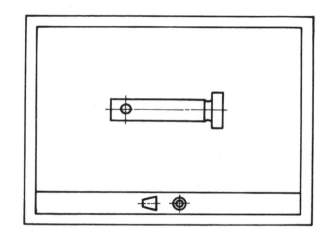

Fig. 1.1 Single-part drawing where one view, with explanatory notes, is sufficient

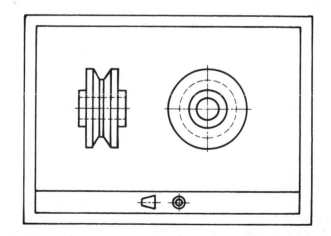

Fig. 1.2 Single-part drawing where two views are required

1

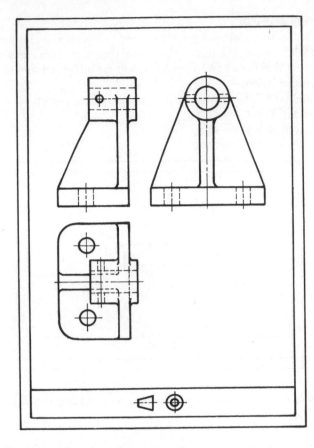

Fig. 1.3 Single-part drawing where three views are required

Assembly and general-arrangement drawings (fig. 1.4)
These drawings show the entire finished product, with all parts assembled together in their correct positions.

These drawings may include overall dimensions, fitting dimensions, and a parts list. They may also include a grid-reference system for quick location of parts comprising the assembly.

Design-layout drawings
Design-layout drawings are usually sketches which include sufficient information for draughtsmen to produce the formal drawings.

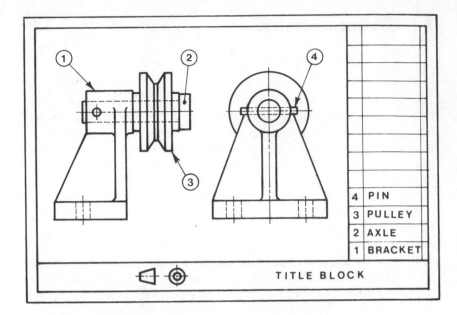

4	PIN
3	PULLEY
2	AXLE
1	BRACKET

TITLE BLOCK

Fig. 1.4 Assembly drawing

Pictorial drawings
Pictorial drawings are usually one-view representations of a component or an 'exploded' view showing a number of components separated but in their correct order for assembly (see page 6).

1.3 Drawing equipment
It is advisable to purchase the best drawing instruments you can afford.

The *drawing board* must be kept clean and smooth, and care should be taken not to damage the edge of the board. The drawing paper should be fixed to the drawing board with clips or adhesive tape – never with drawing pins, as they damage the board and the paper.

The *tee square* is used only for horizontal lines and should be held tightly against the edge of the board when in use. The working edge of the tee square should be bevelled, and care should be taken not to damage it.

Set squares of 45° and 60°/30°, or an adjustable set square with bevel edges, are required.

Pencils   H grades of lead are hard and B grades are soft. The 2H grade is generally used for thin line work, dimensions, centre lines, hidden detail, etc. The H grade is used for thick line work, visible outlines etc. The HB grade is used for lettering, numerals, and sketching.

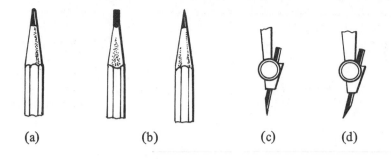

(a)   (b)   (c)   (d)

Fig. 1.5 (a) Pencil sharpened to a cone point

        (b) Two views of a pencil sharpened to a chisel point

        (c) Small spring-bow-compass lead sharpened in one plane on the inside

        (d) Large spring-bow-compass lead sharpened in one plane on the outside

Pencils can be sharpened into a cone point or a chisel point, using a sand-paper block, as shown in fig. 1.5. The advantage of the chisel point is that thick and thin lines can be drawn with the same pencil, the edge is retained longer, and it is less likely to break. A cone will last longer if the pencil is rotated occasionally while drawing or lettering.

Compasses of the spring-bow type with a shouldered pin are preferable. Compass leads should be of the same grade as the drawing pencils used. They should be sharpened in one plane only — on the inside for small compasses and on the outside for large compasses, as shown in fig. 1.5 (c) and (d). The different grades of leads required could be taken from unwanted drawing pencils.

It is desirable to use three sizes of compass: a small spring-bow, a large spring-bow, and a beam compass for very large arcs.

Compasses may be used as *dividers*, if required, by replacing the leads by pins.

Scales should be marked accurately in divisions of 1 mm or, preferably, 0·5 mm over the full length.

A *radius curve* is useful for drawing internal and external fillet radii.

A *French curve* is very useful for drawing curves other than circular curves.

The *eraser* should be a soft white pencil rubber, to ensure that the drawing-paper surface will not be damaged.

An *eraser shield* is very useful for erasing mistakes on drawings without erasing adjacent correct lines.

A *sandpaper block* or a small smooth file is used for sharpening leads.

A *clean duster* is useful for keeping drawing equipment clean.

Hands must always be clean when drawing.

1.4 Layout of drawings

Drawing sheets

Figure 1.6 shows the 'A' series of drawing sheets which are normally used:

    A4   210 mm x 297 mm

    A3   297 mm x 420 mm

    A2   420 mm x 594 mm

    A1   594 mm x 841 mm

    A0   841 mm x 1189 mm

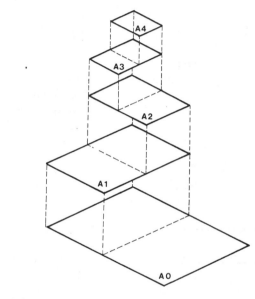

Fig. 1.6 The 'A' series of drawing sheets

Title block

The title block should be at the bottom of the sheet, with the drawing number in the lower right-hand corner. For storage reference purposes, the drawing number may also appear elsewhere on the drawing.

The following basic information should be included in a title block:

a) name of firm (or college),

b) name of draughtsman (or student),

c) title (name of components),

d) date,

e) projection symbol (see pages 27 and 31),

f) scale ratio (see page 45),

g) drawing number.

3

1.5 Lines

All lines should be black, bold, and of consistent density and thickness.

Two thicknesses of line are recommended: a thin line and a thick line two to three times thicker than the thin line.

Dashed lines should be of consistent length and spacing.

Type		Description	Application
A	———————	Thick continuous	Visible outlines and edges
B	———————	Thin continuous	Dimensions and leader lines, projection lines, hatching, outlines of adjacent parts and revolved sections
C	∼∼∼∼∼	Thin continuous irregular	Limits of partial views or sections when the line is not an axis
D	— — — — —	Thin short dashes	Hidden outlines and edges
E	—— · —— · ——	Thin chain	Centre lines, extreme positions of movable parts
F	▬ — · — · — ▬	Chain, thick at ends and at changes of direction, thin elsewhere	Cutting plane

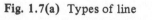

Fig. 1.7(a) Types of line

A visible outline always takes preference over any other type of line, and a thin dashed line showing hidden detail is drawn in preference to a centre line.

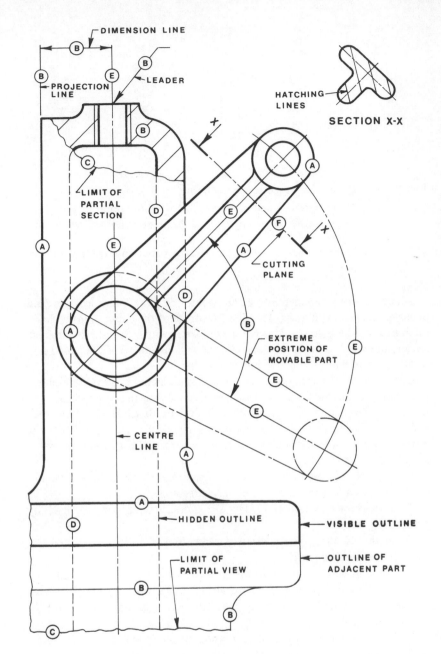

Fig. 1.7(b) Applications of the various types of line

4

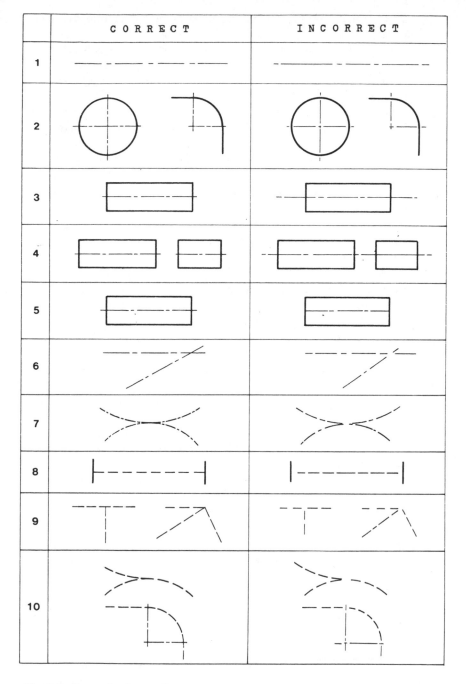

	CORRECT	INCORRECT
1		
2		
3		
4		
5		
6		
7		
8		
9		
10		

Fig. 1.8 General rules on line work

General rules on line work (see fig. 1.8)
1. All chain lines should start and finish with a long dash.
2. Centre lines should cross one another at long-dash portions of the line at the centres of circles, arcs, etc.
3. Centre lines should extend only a short distance beyond the feature, unless required for dimensioning etc.
4. A centre line should not extend through the spaces between views.
5. A centre line should not terminate at another line.
6. Where angles are formed in chain lines, long dashes should meet or cross at the intersections.
7. Arcs should join at tangent points.
8. Dashed lines should start and end with dashes in contact with the hidden or visible outline.
9. Dashed lines should meet or cross with dashes at the intersection.
10. If a dashed line meets a curved line tangentially, it should be with solid portions of the lines.

1.6 Lettering and numerals

1. Capital letters should be used.
2. All letters should be of the same height, using the full spacing of the guidelines.
3. There should be equal spacing between letters, and the spacing between words should not be less than a letter width.
4. All lines should be bold and uniform.
5. Dimensions and notes should be not less than 3 mm tall, but titles and drawing numbers should be larger.
6. All notes should be placed in a horizontal position.
7. Underlining of notes is not recommended – larger letters may be used instead.

1.7 Technical terms

It is necessary for engineers to know and understand the technical terms describing components and their features.

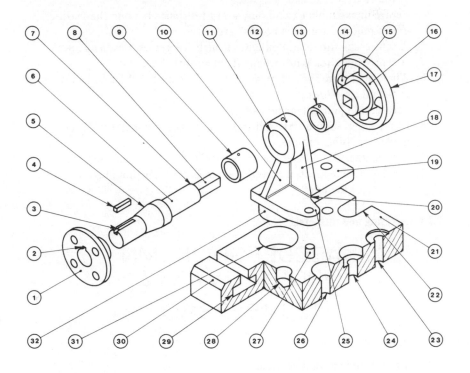

1. A *flange* is a projecting thin disc on pipes or couplings joining two shaft ends together.
2. and 3. A *keyway* is a groove in a shaft or a hub machined to accommodate a corresponding key.
4. A *key* is a piece of shaped metal which is inserted in a shaft and a hub to prevent relative movement between those two parts.
5. A *taper* is a gradual change in diameter of a component along its length.
6. A *shaft* is a cylindrical rotating rod upon which parts are fixed, used for transmission of motion.
7. A *shoulder* is a sudden change in diameter.
8. A *square on a shaft* is a length of the shaft with a square cross-section.
9. A *bush* is a plain bearing supporting a rotating shaft and can be easily replaced when worn out.
10. A *web* is a thin flat part connecting heavier parts of a component.
11. A *bore* is a cylindrical hole along a tube or a boss.

12. A *boss* is an enlarged protruding round part of a casting, used to accommodate a hole.
13. A *collar* is a separate ring of rectangular section or an integral part of a shaft used for axial location.
14. *Spokes* are rods radiating from the hub to the rim of a wheel.
15. A *rim* is the outer part of a wheel.
16. A *hub* is the inner part of a wheel.
17. A *pulley* is a small wheel with a flat or grooved rim to carry a belt, rope, etc.
18. A *rib* is a thin part used to support or strengthen heavier parts of a component.
19. A *bracket base* is the bottom part of a projecting support, usually fixed to a flat surface.
20. A *fillet* is an internal corner of a casting etc. which is curved to assist the flow of molten metal during casting and also to make the corner stronger by reducing stress concentrations.
21. A *table* is the flat top on which working components can be fixed.
22. A *slot* is an elongated hole or groove.
23. A *spot-faced* surface is a flat circular surface concentric with a hole, used for seating screw heads etc.
24. A *counterbored hole* is a hole, part of which is of larger diameter and flat-bottomed to conceal screw heads etc.
25. A *lug* is a projection from a casting etc., used for fastening and adjusting purposes.
26. A *countersunk hole* is a hole, part of which is conical to receive screw heads.
27. A *dowel* is a headless cylindrical pin used for precise-location purposes.
28. A *blind-drilled hole* is a hole which does not pass completely through the component.
29. A *tee groove* or *tee slot* is a long aperture used to accommodate fixing bolts, preventing them turning.
30. A *chamfer* is a surface produced by bevelling square edges.
31. A *recess* is a shallow hole to suit the shape of a spigot or a similar matching part.
32. A *spigot* is a projection which fits into a corresponding recess and is used for precise-location purposes.

1.8 Abbreviations and symbols

There are a number of common engineering terms and expressions which are frequently replaced by abbreviations or symbols on drawings, to save space and draughting time. Some of the abbreviations and symbols recommended by the British Standards Institution in BS 308 are illustrated below and listed on the right.

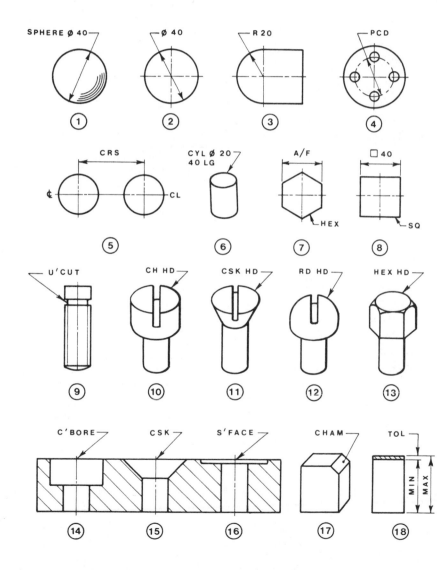

No.	Abbreviations or symbol	Term
1	SPHERE Ø (preceding a dimension)	Spherical diameter
2	Ø (preceding a dimension)	Diameter
3	R (preceding a dimension)	Radius
4	PCD	Pitch-circle diameter
5	CRS	Centres
	CL or ₵	Centre line
6	CYL	Cylinder or cylindrical
	LG	Long
7	A/F	Across flats
	HEX	Hexagon or hexagonal
8	☐ (preceding a dimension)	Square
	SQ (in a note)	Square
9	U'CUT	Undercut
10	CH HD	Cheese head
11	CSK HD	Countersunk head
12	RD HD	Round head
13	HEX HD	Hexagonal head
14	C'BORE	Counterbore
15	CSK	Countersunk
16	S'FACE	Spot face
17	CHAM	Chamfered
18	TOL	Tolerance
	MAX	Maximum
	MIN	Minimum
—	ASSY	Assembly
—	DIA (in a note)	Diameter
—	DRG	Drawing
—	EXT	External
—	FIG. (with full stop)	Figure
—	INT	Internal
—	LH	Left-hand
—	MATL	Material
—	NO. (with full stop)	Number
—	RH	Right-hand
—	SCR	Screwed
—	SH	Sheet
—	SK	Sketch
—	SPEC	Specification
—	STD	Standard
—	▷	Taper on diameter or width

1.9 Conventional representation of common features

There are many common engineering features which it is difficult and tedious to draw in full. In order to save draughting time and space on drawings, these features are represented in a simple conventional form as shown in fig. 1.9.

External screw threads

The crests of the male thread of a stud are defined by a continuous thick line, and the roots of threads by a parallel continuous thin line. The distance between these parallel lines should be approximately equal to the depth of thread, i.e. approximately one eighth of the major diameter of the thread (see page 77).

The limit of the useful length of the thread – 'full thread' – is shown by a continuous thick line (see page 77).

In an end view, the thread roots inside the material are represented by an inner thin broken circle.

Internal screw threads

The tapped hole initially is drilled, which is indicated by the thick outlines. When the hole is tapped, the roots of the threads are defined by a parallel continuous thin line.

In an end view, the thread roots inside the material are represented by an outer thin broken circle.

In a sectional view, the hatching lines are drawn across the thin lines (see pages 38 and 77).

A screw-thread assembly

The male thread of an inserted stud takes precedence over the female thread of the hole.

The hatching lines are not drawn across the thick lines.

In an end view, the male part which is nearest to the observer is represented.

Interrupted views

To save space, it is permissible to show only those parts of a long component which are sufficient for its definition. All break lines are thin and continuous. tinuous.

A square on a shaft

To avoid drawing an additional view, a square or a flat on a round part may be indicated by two diagonal continuous thin lines.

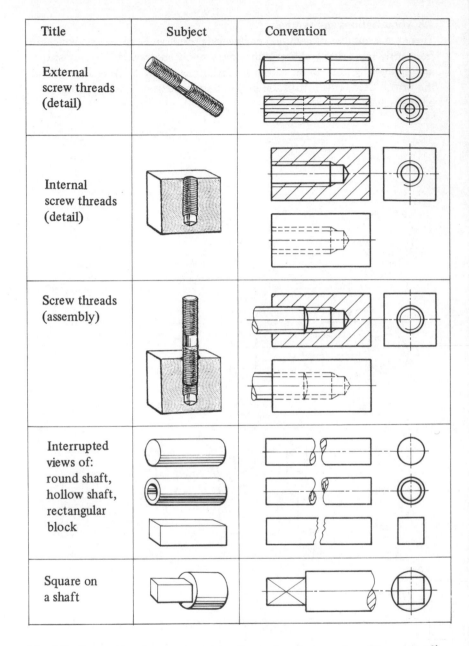

Title	Subject	Convention
External screw threads (detail)		
Internal screw threads (detail)		
Screw threads (assembly)		
Interrupted views of: round shaft, hollow shaft, rectangular block		
Square on a shaft		

Fig. 1.9 Conventional representation of common features (cont'd on page 9)

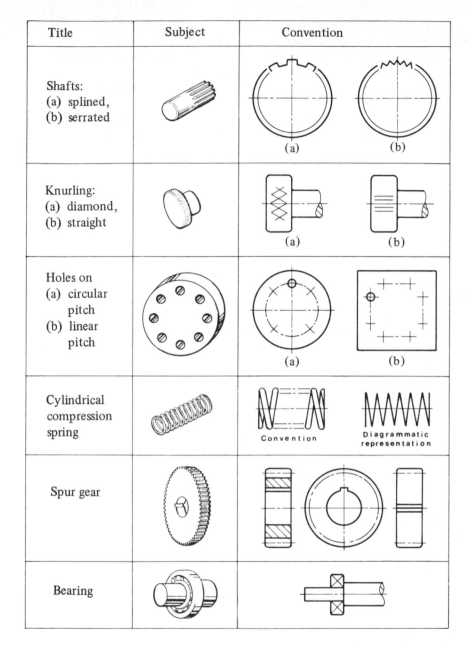

Title	Subject	Convention	
Shafts: (a) splined, (b) serrated		(a)	(b)
Knurling: (a) diamond, (b) straight		(a)	(b)
Holes on (a) circular pitch (b) linear pitch		(a)	(b)
Cylindrical compression spring		Convention	Diagrammatic representation
Spur gear			
Bearing			

Fig. 1.9 Conventional representation of common features (*cont'd*)

Splined and serrated shafts
Only a few splines or serrations need be shown, with the root circle represented by an inner thin circle.

Diamond and straight knurling
Knurling provides a rough surface to facilitate the operation of the component by hand. (A spring-bow compass usually incorporates both types of knurling.)

Holes on circular and linear pitch
When a large number of repeated holes are required, it is sufficient to draw only one hole and to indicate the position of the remainder.

Cylindrical compression springs
Coils are drawn only at each end of the spring and are connected by parallel thin chain lines indicating the repetition of coils.

Spur gears
The front view of a spur gear consists of the outside thick circle and the pitch 'centre-line' circle.

In the sectional end view the gear teeth are not sectioned, but the centre line and the root of the teeth are shown.

In the remaining end view the centre lines are shown and the direction of teeth is indicated by thick parallel lines.

Bearings
Sometimes it is necessary to draw sectional views of a number of ball and roller bearings on one drawing. The complicated sectional views can be replaced by a conventional representation consisting of the bearing outline and thin-line diagonals.

2 Organisation

2.1 Stages in the development of a new product

Initially the customer's requirements are considered by a designer, to produce specifications which take into consideration all the different factors that influence the design of a product. At this stage, various solutions to the design problems are considered and the best ones are selected.

The final design is eventually produced in the form of a design layout by the designer and passed on to his draughtsmen.

A general-assembly drawing and sub-assembly drawings are prepared by the design draughtsmen, the production specifications are drawn up, and finally the single-part (component) drawings are produced by the detail draughtsmen.

After drawings have been checked, traced, and printed, copies are sent to the manufacturing department, where the designed product is made, assembled, and tested.

2.2 Standard parts

Before national and international standards came into being, there was no uniformity of products manufactured by different manufacturers. Nowadays, if two mating standard parts are ordered from two different suppliers, providing the specification is the same in both cases, those two parts will fit together when assembled.

In Great Britain, this control of variety, or standardisation, is administered by the British Standards Institution, which is a member of the International Organisation for Standardisation (ISO).

The standards used in a typical drawing office cover many topics: terminology, definitions, symbols, preferred numbers and sizes, materials, tools, equipment, papers, processes, practices, safety, standard parts, etc.

The use of standard parts

a) simplifies the design, as standard parts are usually already designed and manufactured;

b) makes production more economical, as standard parts are mass-produced, hence relatively cheaper, and are usually kept in stock;

c) reduces the cost of maintenance of a product already in use, due to the interchangeability of standard parts.

2.3 The drawing office

The functions of a drawing office vary from firm to firm. A relatively large firm may have a separate design office, whereas the majority of firms have drawing offices incorporating the design section.

The main functions of a typical drawing office are

a) to prepare the design layouts and assembly and production drawings necessary for the manufacture of products;

b) to make decisions on materials to be used, methods of manufacture, heat treatment, etc.;

c) to calculate the stresses for the designed components, to ensure that the components will withstand the applied forces when manufactured and in use;

d) to estimate from the drawings the cost of manufacture of designed components;

e) to store all drawings, technical information, and reference material;

f) to provide a technical service for all departments in the firm; and

g) to liaise with people outside the firm.

2.4 The print room

The function of a reproduction or print room is to copy engineering drawings, tracings, and documents. The copies obtained are used for reference, manufacturing, assembly, and storage purposes.

There are three popular reproduction methods: the dye-line (diazo) process, the microfilm process, and the xerographic process.

The dye-line (diazo) process

The tracing is placed over special printing paper and is exposed to ultra-violet light, which bleaches away the sensitive coating on the printing paper, except where the ink lines on the tracing prevent light passing. The exposed paper is then developed, to show the lines of the tracing in dark colours.

The print obtained is the same size as the original drawing.

The microfilm process

The original drawing or tracing is photographically reduced on to a film. The film, when developed, may be stored on a reel or may be cut and placed in cellophane envelopes or mounted in cards or frames. When the drawing is to be referred to, the film is projected on to a screen or enlarged and printed directly.

The original drawing or tracing must be of good standard – the lines must be drawn black and thick and well spaced, detail reduced, and lettering clear.

Xerographic process

This uses a plate coated with a material which conducts electricity when exposed to light and acts as an insulator in the dark. The plate surface is positively charged in the dark and a drawing is projected on to it. Plate areas corresponding to white areas of the drawing are exposed to light and lose their charge; dark areas, corresponding to the image of the drawing, retain their positive charge.

The plate is then dusted with a negatively charged powder which adheres only to the positively charged areas. A positively charged paper is pressed against the plate, attracting the particles of powder, and finally heat is applied to fuse those particles permanently to the paper.

This process is suitable only for reproducing small-size drawings and documents.

2.5 Drawing-office personnel

Designer

A designer is usually a professional engineer, a technologist with a degree. He or she must be creative — be able to develop ideas and solve design problems. A designer's technical knowledge should include such disciplines as mathematics, theory of machines, strength of materials, fluid mechanics, thermodynamics, materials science, production methods, electricity, electronics, ergonomics, aesthetics, etc.

He or she must be able to convey ideas and instructions clearly, accurately, and concisely, to guide the draughtsmen.

A designer's main functions are

a) to originate and finalise designs for a proposed product to satisfy the function, cost, manufacture, and market requirements in consultation with production engineers, industrial (artistic) designers, purchasing specialists, customers, and other interested persons;

b) to advise the drawing-office staff on technical difficulties;

c) to ensure that the product, after it has been manufactured, is functional, reliable, and easily maintained.

A designer may hold the post of chief draughtsman or drawing-office manager.

Chief draughtsman

The chief draughtsman is a professional or technician engineer with a higher technician diploma or certificate. His or her functions as drawing-office manager are

a) to organise, direct, and co-ordinate the work of the drawing office and its resources;

b) to determine staffing, promotion, and necessary training for junior staff;

c) to co-ordinate the activities of drawing-office staff — section leaders, design and detail draughtsmen, checkers, tracers, etc.;

d) to advise staff on technical problems;

e) to liaise with other departments inside the firm and with people outside the firm;

f) to examine design specifications provided by a designer;

g) to instruct section leaders on the distribution and detailing of work.

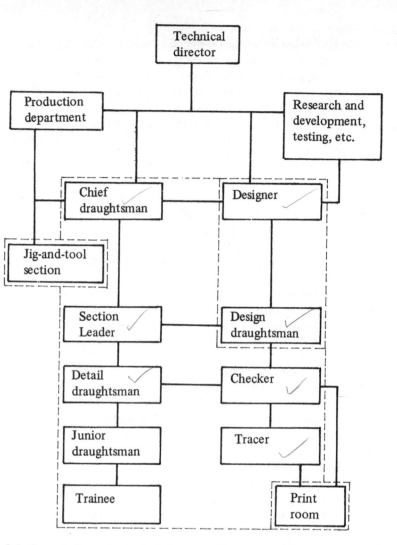

Fig. 2.1 Organisation of a typical drawing office and design office

Section leader

The section leader is usually a technician engineer supervising a small team of draughtsmen and is directly responsible to the chief draughtsman. His or her main functions are

a) to distribute work,

b) to make sure that work is satisfactory and completed on time,

c) to arrange for completed drawings to be checked, traced, reproduced, distributed, and stored.

11

Detail designer or design draughtsman

A design draughtsman is usually a technician engineer with a higher technician certificate or diploma. His or her main functions are

a) to prepare the general-assembly drawings from a design layout;
b) to calculate the main dimensions and masses of components, if required;
c) to decide on tolerance fits, threads, etc.;
d) to consider materials to be used and methods of manufacture;
e) to prepare working specifications for detail draughtsmen, indicating overall measurements and functional requirements.

Draughtsman

A draughtsman is usually a technician with an ordinary technician certificate or diploma. His or her main functions are

a) to prepare detailed drawings of individual components, parts, and sections;
b) to insert on drawings production guidance and information, tolerances, materials to be used, surface finish, heat treatment, etc.;
c) to calculate detail dimensions;
d) to prepare a parts list;
e) to arrange for completed drawings to be checked, traced, and reproduced for use as working drawings in the production department.

Checker

The checker is usually a technician, and his or her main functions are

a) to check finished engineering drawings to ensure that all specification requirements have been met, all dimensions and tolerances have been shown, and all information has been inserted regarding manufacture, materials to be used, surface finish, heat treatment, etc.;
b) to check the accuracy of dimensions;
c) to ensure that safety has been considered;
d) to make sure that all parts will function correctly and are easy to manufacture.

Finally, the checker refers drawings back to the draughtsmen for ratification.

Tracer

The tracer's main functions are

a) to make tracings, usually in ink, of the original drawings;
b) to insert additional detail, if instructed;
c) to refer drawings to the print room and then to file drawings, tracings, and copies of drawings if required.

Drawings are traced in order to improve the quality of prints and also to make a more durable record of original drawings.

Every draughtsman and tracer should know about the different methods of drawing reproduction, in order to get the best possible prints from his or her drawings.

2.6 Test questions on chapters 1 and 2

1. State what is meant by engineering drawing.
2. Give reasons for using a drawing in preference to written and spoken communication in engineering.
3. Drawings should convey three types of general information. Name two of them.
4. State the function of a single-part drawing and list the items it should include.
5. State the function of an assembly drawing.
6. Name five items of basic information that a drawing title block should include.
7. State why the following types of drawing are needed:
   a) single-part or component,
   b) sub-assembly,
   c) general-assembly.
8. If a thick continuous line is used to represent the visible outlines and edges of a component on a drawing, what will the following lines represent:
   a) a thin continuous line,
   b) thin short dashes,
   c) a thin chain line.
9. Which of the following statements are true and which are false?
   a) A visible outline takes preference over any other line on a drawing.
   b) A centre line takes preference over a thin dashed line showing hidden detail.
   c) Centre lines should cross one another at long-dash portions of the lines.
   d) Underlining of notes is not recommended.
10. With the help of simple sketches, identify and explain the following abbreviations or symbols.

CRS	A/F	U'CUT	HEX	HD	TOL	LG
S'FACE	CHAM	□	PCD	CL	STD	RH

11. Explain briefly why standard abbreviations are used on engineering drawings.
12. With the help of sketches, show the conventional representation of the following:
    a) an external thread,
    b) an interrupted view of a hollow shaft,
    c) a square on a shaft,
    d) diamond knurling,
    e) a splined shaft,
    f) a bearing.

13. Put the following stages in the development of a new product in their correct order, naming the personnel involved and their functions: (a) single-part drawing, (b) manufacture of a product, (c) printed copy of a drawing, (d) assembly of a product, (e) design layout, (f) assembly drawing.

14. State four main functions of a typical drawing office.

15. Indicate the reasons for copying engineering drawings in a print room.

16. Describe very briefly two popular drawing-reproduction processes.

17. Describe at least three main functions of each of the following drawing-office personnel:
    a) the chief draughtsman,
    b) the designer,
    c) the draughtsman,
    d) the checker,
    e) the tracer.

18. State three reasons for using standard parts in engineering.

19. With the help of a simple block diagram, indicate the organisation of a drawing office.

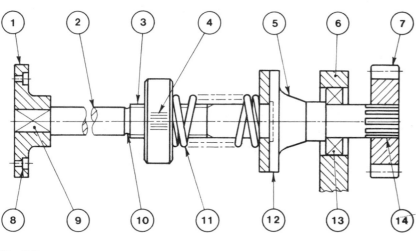

Fig. 2.2

20. Study fig. 2.2 and identify the referenced items with the help of the following descriptions: (a) square on shaft, (b) counterbored hole, (c) interrupted view, (d) bearing, (e) external thread, (f) undercut, (g) straight knurling, (h) splined shaft, (i) compression spring, (j) taper, (k) bearing housing, (l) shaft coupling, (m) spur gear, (n) shaft flange.

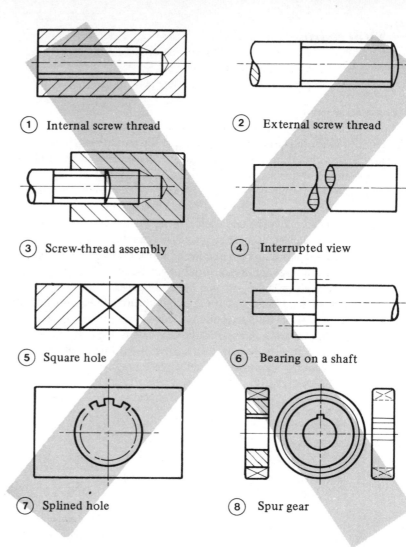

1. Internal screw thread
2. External screw thread
3. Screw-thread assembly
4. Interrupted view
5. Square hole
6. Bearing on a shaft
7. Splined hole
8. Spur gear

Fig. 2.3 *Incorrect* conventional representations

21. Figure 2.3 shows six conventional representations of common features, which are *not* drawn in accordance with BS 308. Redraw each item correcting all deliberate mistakes. Items 1, 2, 4, 5, 6, and 7 include at least three mistakes; items 3 and 8 include at least six mistakes. Tracing paper may be used.

3 Sketching

3.1 Freehand sketching

The importance of freehand sketching is very often underestimated.

The ability to sketch quickly, accurately, and in good proportion is essential to engineering communication. The freehand technique should be employed by an engineer as a better means of visualising problems and quickly organising his or her thoughts to avoid wasting time on more formal drawing methods.

The designer nearly always sketches his or her first ideas in pictorial form, as they are easily visualised before the proper drawings are produced.

A freehand sketch is a drawing in which all proportions and lengths are judged by eye and all lines are drawn without the use of drawing instruments – the only tools used are pencil, eraser, and paper.

The ability to sketch is a skill which is acquired through learning initially to draw freehand vertical and horizontal straight lines, squares, circles, ellipses, and curves. Circular curves must be drawn with the ball of the hand inside the curve, and straight lines must be drawn by resting the weight of the hand on the backs of the fingers, as shown in fig. 3.1.

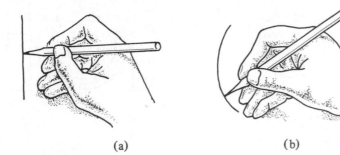

(a)   (b)

Fig. 3.1

To sketch a straight line (fig. 3.1(a))

1. Mark the end points of the required line.
2. Sketch a light trial line using several short strokes, with the eye fixed on the point towards which the straight line is being drawn.
3. Finally, press the pencil down to get a uniform bold straight line.

To sketch a circle (fig. 3.2)

1. Sketch centre lines and the enclosing construction square, then sketch the diagonals and step off distances from the centre equal to the radius.
2. With the hand positioned within the circle and pivoted at the wrist, sketch the trial circle consisting of eight short arcs.
3. Finally, press the pencil down to get a uniform bold line and erase all construction lines as required.

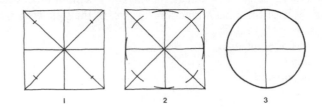

Fig. 3.2 Three stages in sketching a circle

Circles may also be drawn by rotating the paper with the left hand about one of the fingers of the right hand acting as a pivot, while the right hand holds the pencil at the required length. Figure 3.3 shows how to hold the pencil for sketching large and small circles.

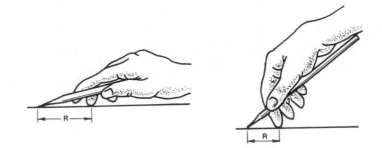

Fig. 3.3 Sketching large and small circles

Alternatively, circles may be sketched with the help of a piece of paper on which the radius is marked. With one mark kept on the circle centre, the other mark is used to plot all the required points.

Form and proportion

The required sketching skills can be achieved by constant practice with real-life objects of different shapes, special attention being paid to density of line, good form, and relative proportions.

To obtain good form and proportion, a light construction framework of rectangular boxes, cubes, cylinders, cones, etc. can be used to represent the outlines of the objects sketched, as shown in fig. 3.4.

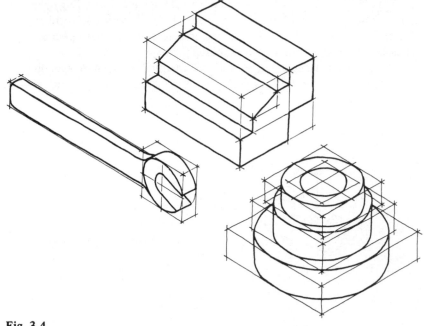

Fig. 3.4

To determine the required proportions, the pencil should be held at arm's length, marking the height of the object with the tip of the thumbnail, fig. 3.5. The arm then is rotated until the pencil coincides with another edge of the object and then an estimate is made along the pencil of the ratio of the two lengths.

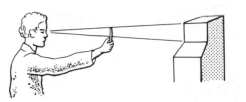

Fig. 3.5

4 Pictorial projection

Projection is a method of representing visually a three-dimensional object on two-dimensional drawing paper. A pictorial projection is a method of producing a two-dimensional view of a three-dimensional object that shows three main faces indicating the height, width, and depth simultaneously, as in fig. 4.1.

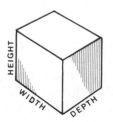

Fig. 4.1

4.1 Perspective projection
Appreciation of perspective is essential for learning the fundamentals of freehand sketching. Objects at a distance appear to be smaller than those which are near. Two parallel lines representing the edges of a straight road seem to come closer together and then meet at a point on the horizon. That point is called the vanishing point (VP), fig. 4.2.

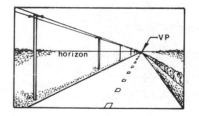

Fig. 4.2

Perspective projection involves a number of receding lines called projectors converging at one, two, or more vanishing points. The objects sketched are then presented as they would appear when observed from a particular point in real life.

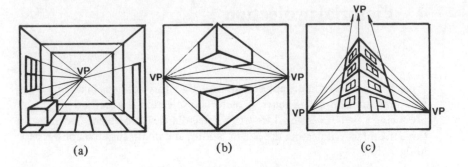

(a)   (b)   (c)

Fig. 4.3 (a) One-point, (b) two point, and (c) three-point perspective

In fig. 4.3(a), one-point perspective is shown, where one of the principal faces is parallel to the picture plane.

In fig. 4.3(b), two-point perspective is shown, where all principal faces are inclined. This method is commonly used for industrial sketching.

In fig. 4.3(c), three-point perspective is shown, where one vanishing point is outside the picture frame.

For engineering purposes when sketching small objects, the vanishing points are considered to be placed outside the frame of the drawing paper. This is because such objects are usually viewed from a close distance.

In fig. 4.4, the features of the objects above the eye-line or horizon are seen from below, and the features below the horizon are seen from above.

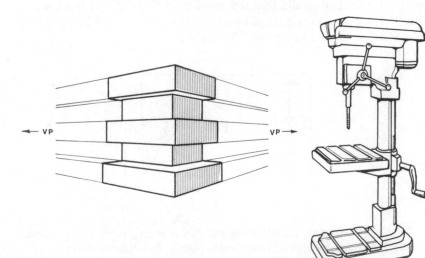

Fig. 4.4

For guided perspective sketching, use the grid in fig. 4.8. Place tracing paper over the grid and sketch the objects required using the projectors. Alternatively, draw your own projectors radiating from two vanishing points on the horizon.

Figure 4.5 shows a block with a number of holes and protruding cylinders drawn in perspective projection. The perspective grid was used for this guided drawing. Note how all axes, block edges, and sides of cylinders either are vertical or converge towards two vanishing points on the horizon.

All ellipses representing circles were drawn at 90° to the corresponding longitudinal axes or centre lines. (All 90° angles on the drawing are represented by small squares.)

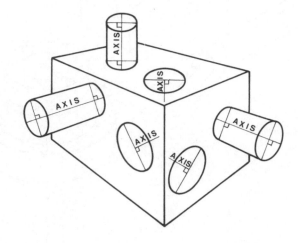

Fig. 4.5

4.2 Test questions

1. The detail shown in fig. 4.6 requires holes to be drilled at the positions marked + and cylindrical (dowel) pins to be inserted at the positions marked *.

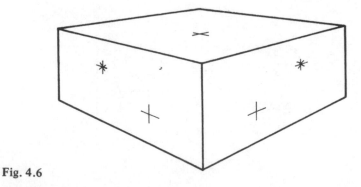

Fig. 4.6

16

Redraw or trace the block and then sketch the holes and pins. The holes are to be approximately 15 mm diameter and the dowels approximately 10 mm diameter, protruding 20 mm from the block. Hole and pin axes are to be normal (at 90°) to the block surfaces.

2. Figure 4.7 shows the outlines of tools that are used by engineering workers of different trades: (1) hammer, (2) double-ended spanner, (3) pliers, (4) adjustable spanner, (5) brace, (6) G-clamp, (7) hand saw, (8) tool-maker's clamp, (9) blow lamp, (10) bench vice, and (11) micrometer.

Sketch freehand each tool in perspective projection, using the construction-box method.

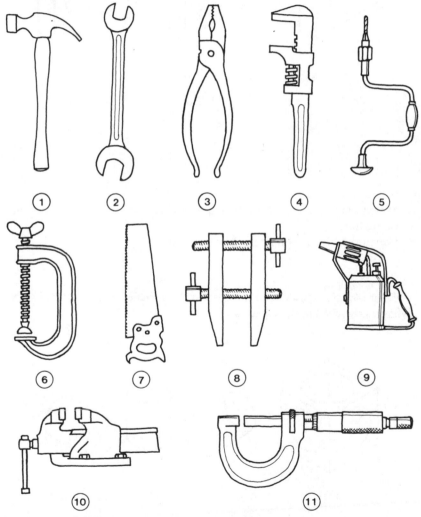

Fig. 4.7

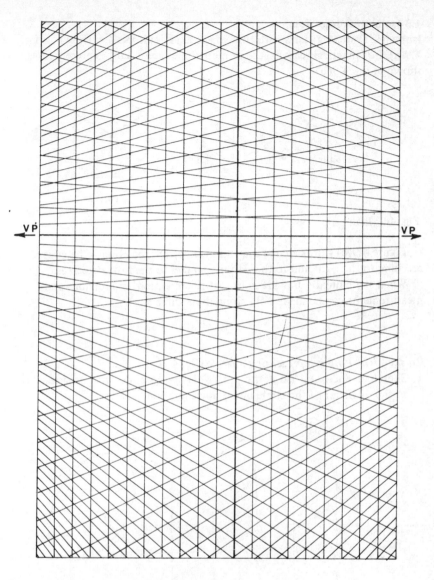

Fig. 4.8 Perspective-projection grid for guided sketching

To sketch an object as seen from above, place tracing paper or thin office paper over the grid below the horizon line.

To sketch the object as seen from below, use the top part of the grid.

4.3 Isometric projection

Isometric sketching starts with three basic axes equispaced as shown in fig. 4.9(a). For practical reasons, the isometric axes are usually represented as shown in fig. 4.9(b).

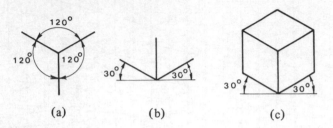

Fig. 4.9 Isometric axes

Figure 4.9(c) shows a cube drawn in isometric projection – the edges receding to the right and to the left are parallel to the isometric axes.

When sketching or drawing in isometric projection, proportions and measurements can be made only along these three axes.

To sketch an ellipse representing a circle (fig. 4.10)

1. Sketch an enclosing 'isometric square', i.e. a rhombus, with its sides equal to the diameter of the circle under consideration.

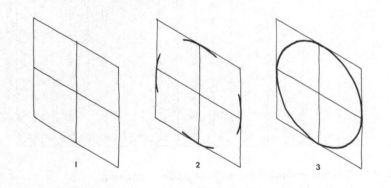

Fig. 4.10

2. Sketch bisecting lines and, at the intersection points, sketch short tangential arcs.
3. Finish the ellipse with a uniform bold line.

Alternative quick method of sketching ellipses (fig. 4.11)

1. Sketch a faint construction line AB representing the longitudinal centre line of a hole or cylinder.

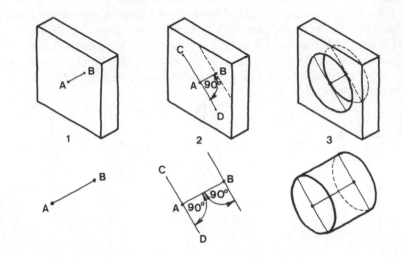

Fig. 4.11

2. Sketch the major axis CD of the required ellipse at 90° to the longitudinal centre line.
3. Sketch the ellipse, estimating the minor axis.

Figure 4.12(a) shows holes sketched in three planes. Figure 4.12(b) shows cylinders sketched in three planes.

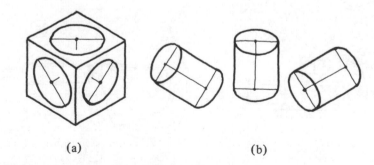

Fig. 4.12

To sketch cylindrical objects, first draw the complete construction ellipses, then draw tangential blending lines as shown in fig. 4.13.

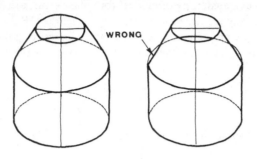

Fig. 4.13

To sketch a nut (fig. 4.14)

1. Sketch three vertical construction lines, placed at D and $D/2$ apart, where dimension D represents a nominal (major) thread diameter. Then sketch a hexagonal prism of height D, using the three vertical construction lines with all receding sides sloping at 45°.

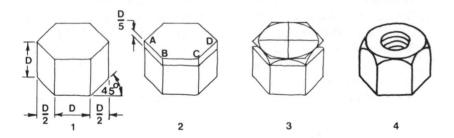

Fig. 4.14

2. Sketch lines parallel to the edges AB, BC, and CD at a distance of $D/5$ from them, and sketch three arcs between these construction lines.
3. Sketch a horizontal tangential ellipse in the upper hexagon, representing the chamfer circle.
4. Sketch two blending arcs and include the threaded hole, which should be slightly smaller than D in diameter. Finish the sketch with uniform bold lines. The threaded hole should be represented by equispaced parallel sections of ellipses.

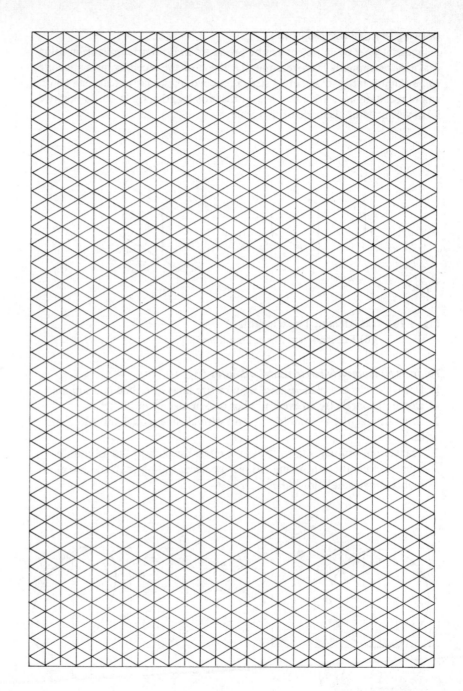

Fig. 4.15   Isometric grid for guided sketching

Drawing in isometric projection

A pictorial view of an object can be produced in isometric projection using drawing instruments. To overcome the effect of the receding lines appearing to be slightly larger than actual size, a reduced or isometric scale can be used, where receding lines are about 0.8 of the true lengths, fig. 4.16. In practice, however, little use is made of this scale.

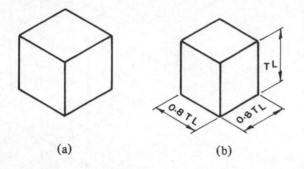

(a)   (b)

Fig. 4.16   (a) All actual true lengths
                 (b) Foreshortened receding lines

To draw an isometric circle using ordinates (fig. 4.17)

1. Draw a circle as a plane figure, with centre lines inscribed in a square. Divide the circle by an even number of equidistant ordinates.
2. Draw the required 'isometric square' with all sides equal and add centre lines.
3. Transfer all ordinates from the plane-circle drawing to the 'isometric square' along the centre line AB, with corresponding ordinate measurements above and below AB.

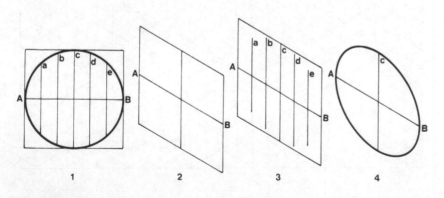

Fig. 4.17

4. Join the plotted points with a uniform bold line, preferably using a French curve to complete the ellipse.

This system of transferring ordinates from plane figures to isometric views may be used for any regular or irregular shape. Figure 4.18 shows irregular shapes in isometric projection, where ordinates representing the uniform thickness of the object are of the same length and are measured along the isometric axes.

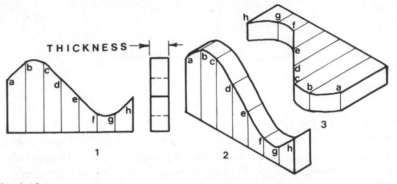

Fig. 4.18

To construct isometric circles using instruments (figs 4.19 and 4.20)

1. Draw an 'isometric square' ABCD, each side being equal, and then draw a long diagonal DB bisecting the two acute angles D and B (fig. 4.19).
2. Join the mid-points of each side to the opposite obtuse angles A and C.
3. Use the intersection points E and F on the diagonal as the centres to draw two small arcs between the nearest mid-points.
4. With centres at A and C, draw the remaining arcs between the mid-points.

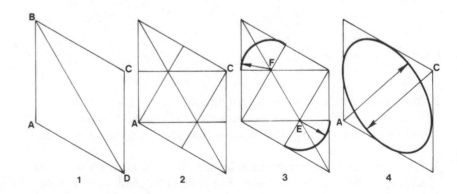

Fig. 4.19

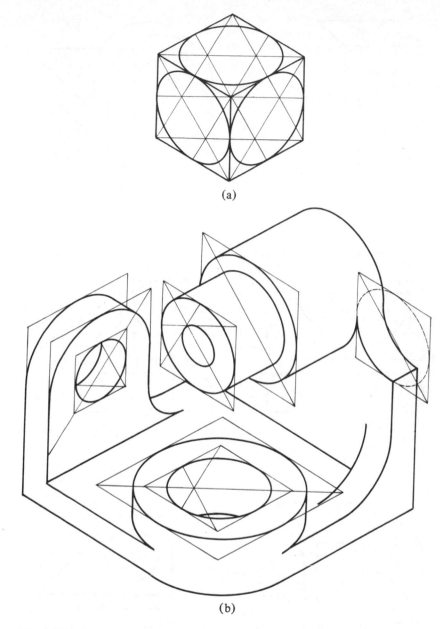

(a)

(b)

Fig. 4.20 Isometric circles in three planes: (a) all construction lines shown, (b) some construction lines omitted for clarity

4.4 Test questions

1. Redraw the components shown in fig. 4.21 in isometric projection, remembering that measurements can be taken only along the basic axes. Each construction square represents a 10 mm measurement and should not be shown on your drawing.

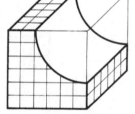

① ② ③

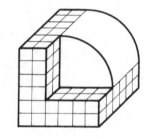

④ ⑤ ⑥

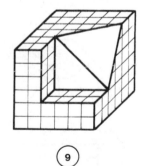

⑦ ⑧ ⑨

Fig. 4.21

2. Redraw in isometric projection the twelve objects shown in fig. 4.22, remembering that measurements can be taken only along the basic axes. Each construction square represents a 10 mm measurement.

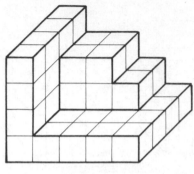

①

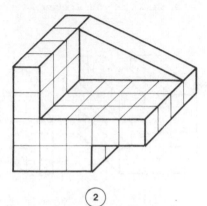

②

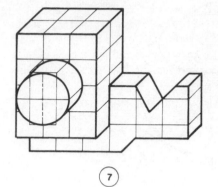

⑦

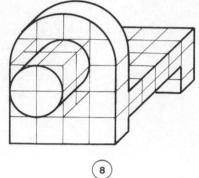

⑧

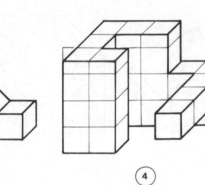

③

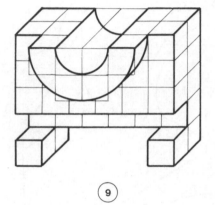

④

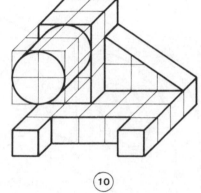

⑨

⑩

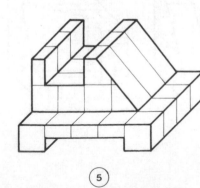

⑤

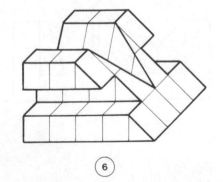

⑥

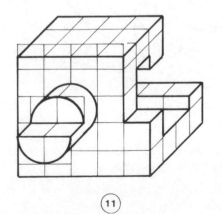

⑩

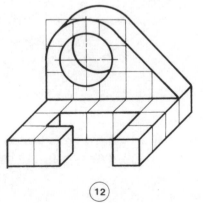

⑪

⑫

Fig. 4.22

22

4.5 Oblique projection

A pictorial view of an object can be produced in oblique projection, where the front face is sketched as a true shape without distortion.

Sketching in this projection is much easier than in isometric projection, since all the circles in the front face are sketched as plane figures instead of as ellipses as in isometric projection.

Oblique sketching starts with two axes — one vertical and one horizontal — together with a third axis which is usually drawn at 45° to the horizontal and along which all measurements are reduced to half true length (TL), as shown in fig. 4.23.

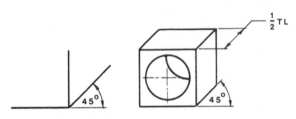

Fig. 4.23   Basic oblique axes

All proportions and measurements can be made only along these three axes.

A comparison of three pictorial projections is shown in fig. 4.24.

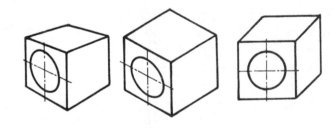

Perspective   Isometric   Oblique

Fig. 4.24   Pictorial projections

Drawing in oblique projection

The main advantage of oblique projection over isometric is that any complicated face — which may be curved, irregular, or have a number of holes — can be drawn as its true shape using drawing instruments.

The receding surfaces can be drawn at any angle but are usually drawn at 45° and are foreshortened to half true length.

All receding surfaces shown in fig. 4.25 are half true length, but at different angles.

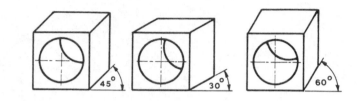

Fig. 4.25

In fig. 4.26 the required semicircles are drawn with centres at A, B, and C positioned along a line at 45° and at distances of half true lengths.

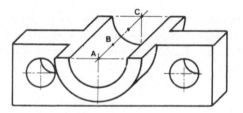

Fig. 4.26

Figure 4.27 indicates the advantages of oblique projection when drawing a complicated cylindrical component.

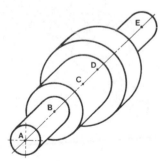

Fig. 4.27

To draw cylindrical shapes in oblique projection (fig. 4.28)

1. Draw the axis of the cylindrical shape at 45° to the horizontal and locate the points A, B, C, and D representing all normal (at 90°) surfaces along the axis at distances of half true length.

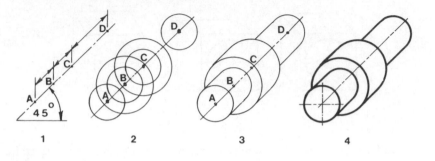

Fig. 4.28

2. Draw the required construction circles at the points A, B, C, and D.
3. Join adjacent pairs of same-size circles by tangents at 45°.
4. To complete, draw all circles and blending tangents with uniform bold lines as shown.

Figure 4.29 shows how complicated cylindrical shapes can be represented in oblique projection.

Fig. 4.29 Crankshaft

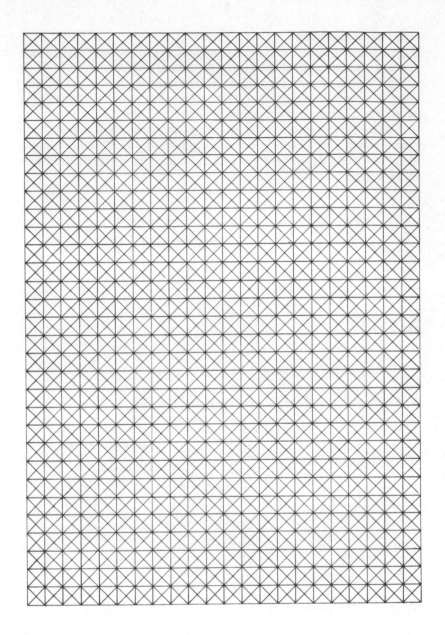

Fig. 4.30 Oblique grid for guided sketching

24

4.6 Test questions

1. Redraw in oblique projection the objects shown in fig. 4.31, remembering that measurements can be taken only along the basic axes. Each construction square represents a 10 mm measurement.

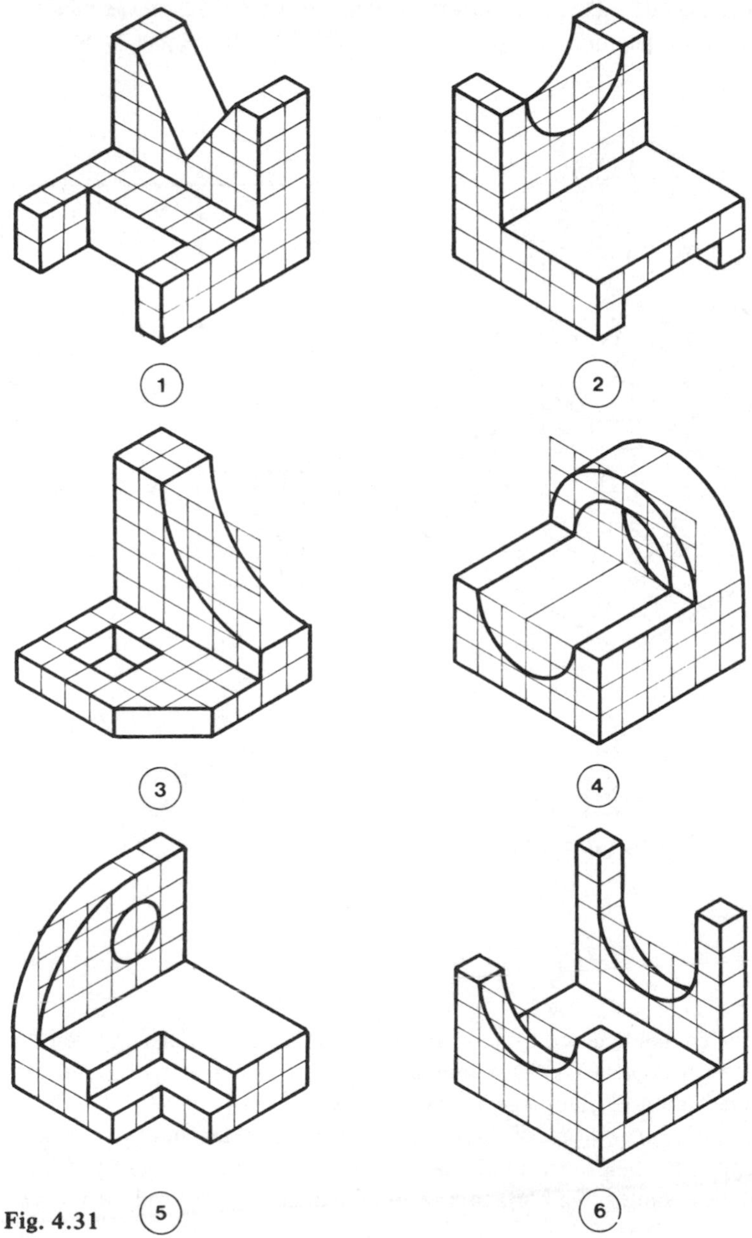

2. Redraw in oblique projection the objects shown in figure 4.32, remembering that measurements can be taken only along the basic axes. Each construction square represents a 10 mm measurement.

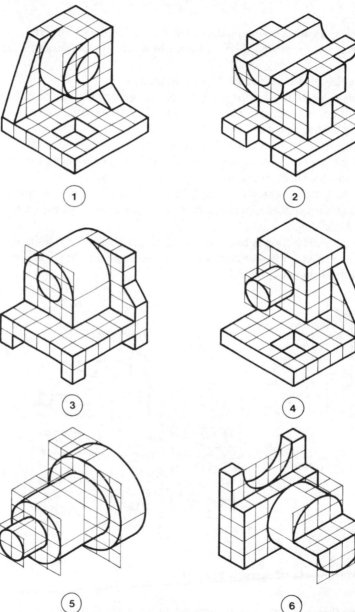

Fig. 4.31

Fig. 4.32

5　Orthographic projection

Orthographic projection is a method of producing a number of separate two-dimensional inter-related views which are mutually at right angles to each other.

Orthographic projection is a multi-view projection.

Using this projection, even the most complex shape can be fully described. This method, however, does not create an immediate three-dimensional visual picture of the object as does pictorial projection.

The ability to visualise or think in three dimensions is essential to the competent reading of drawings and should be developed even before the skills required to produce neat and accurate engineering drawings.

Visualisation is best achieved through the use of models in conjunction with drawings, to promote an understanding of reading drawings and three-dimensional thinking.

Orthographic projection is based on two principal planes — one horizontal (HP) and one vertical (VP) — intersecting each other and forming right angles and quadrants as shown in fig. 5.1.

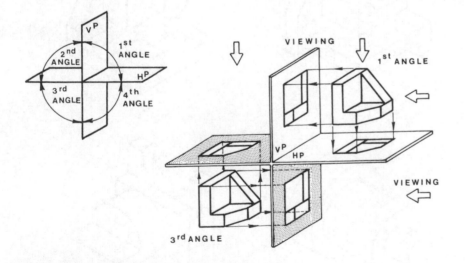

Fig. 5.1 Principle of orthographic projection

Only two forms of orthographic projections are used: first-angle ('European') and third-angle ('American').

5.1 First-angle projection

In first-angle projection, an object is positioned in the space of the first-angle quadrant between two planes, fig. 5.2(a). A view of the object is projected by drawing parallel projecting lines, or projectors, from the object to the vertical principal plane (VP). This view on VP is called a front view. A view similarly projected on to the horizontal principal plane (HP) is called a plan view.

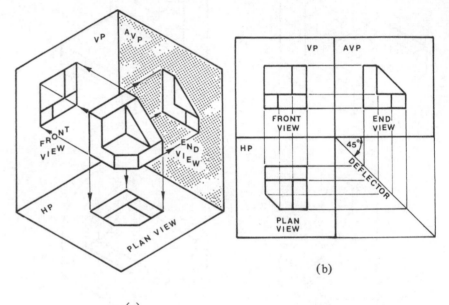

(a)

(b)

Fig. 5.2 First-angle projection

For the complete description of the object, an additional plane, called the auxiliary vertical plane (AVP), is used at 90° to the principal planes, and the view projected on to that plane is called an end view.

By means of projectors, all three planes can be unfolded and three views of the object can be shown simultaneously on drawing paper as in fig. 5.2(b). The end view is projected horizontally and the plan view vertically from the front view.

In first-angle projection, the object always comes between the eye of the observer and the projection plane or view, as shown in fig. 5.3.

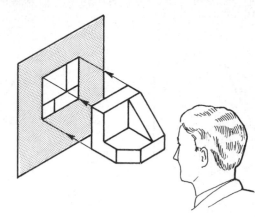

Fig. 5.3 Principle of first-angle projection

The symbol used to indicate first-angle projection is derived from views of a circular taper as shown in fig. 5.4. The symbol shows a front view and a left end view of the circular taper in first-angle projection.

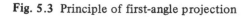

Fig. 5.4 First-angle projection symbol

Sometimes it is necessary to show six views of an object, as in fig. 5.5. To show hidden detail, a thin line with short dashes is used.

As a rule, the minimum number of views should be used, especially to represent simple objects. The views should be selected so that they clearly indicate all the required detail.

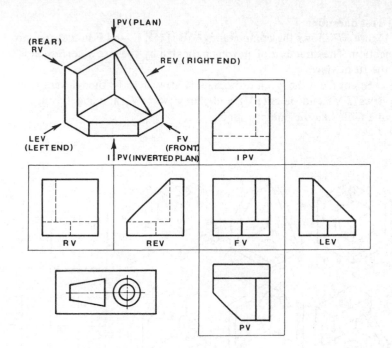

Fig. 5.5 Six views in first-angle projection

In common practice only three views are used: (a) a front view, (b) an end view, and (c) a plan view. These three views are sufficient for a complete description of an object, as shown in fig. 5.6.

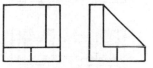

Fig. 5.6 Three views in first-angle projection

5.2 Test questions

1. Figure 5.7 shows the components A, B, C, D, E, and F in pictorial projection. The direction of viewing indicated by the arrow corresponds to the front view.

    Select from the given orthographic views 1 to 18 the relevant front views (FV), end views (EV), and plan views (PV), and insert your answers in a table like the one provided.

2. Draw or sketch, full size, in first-angle projection the components shown in fig. 5.8. Select the views to show most of the features as visible outlines, and include hidden detail where necessary. Each construction square represents a 10 mm measurement.

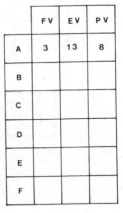

	FV	EV	PV
A	3	13	8
B			
C			
D			
E			
F			

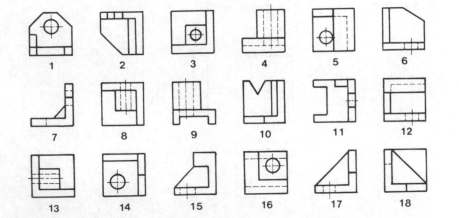

Fig. 5.7

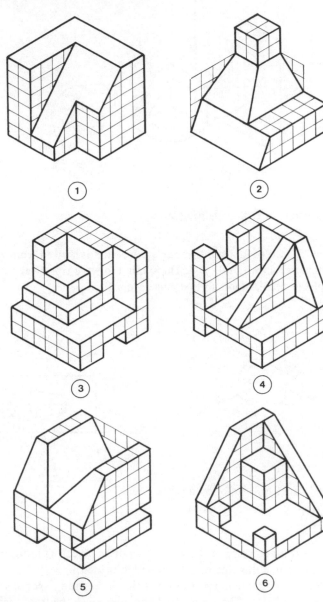

Fig. 5.8

3. Complete the third missing view for each object shown in first-angle
   projection in fig. 5.9. Squared or tracing paper may be used.

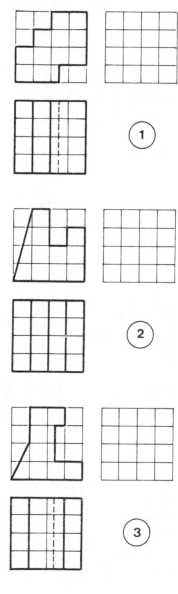
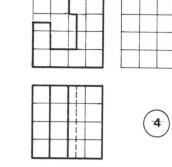
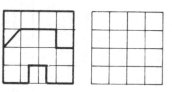
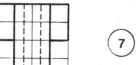
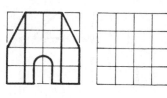
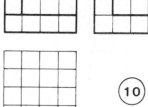

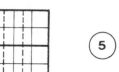
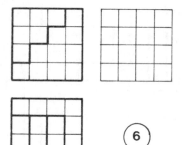
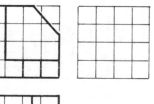

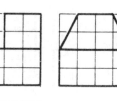

Fig. 5.9

29

4. The components shown in fig. 5.10 are drawn in half-size in first-angle projection. Sketch or redraw each component

a) in isometric projection,

b) in oblique projection, where the position of component 11 may be rearranged.

Isometric and oblique grids may be used (pages 19 and 24).

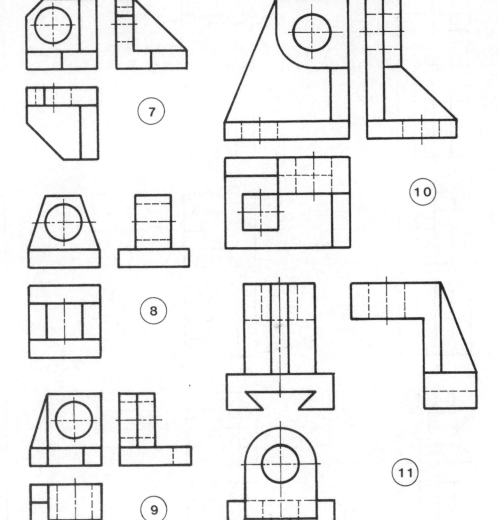
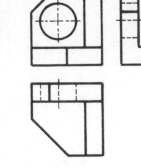
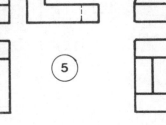

Fig. 5.10

5.3 Third-angle projection

In third-angle projection, an object is positioned in the space of the third-angle quadrant between two principal planes. The planes are imagined to be transparent, and the projected views of the object are viewed through the planes as shown in fig. 5.12(a).

By means of projectors, all three planes of the 'glass box' can be unfolded and three views of the object can be shown simultaneously on drawing paper as in fig. 5.12(b).

In both first- and third-angle projection the views are identical, but the positioning of each is different.

In third-angle projection, the 'transparent' projection plane or view always comes between the eye of the observer and the object, as shown in fig. 5.13.

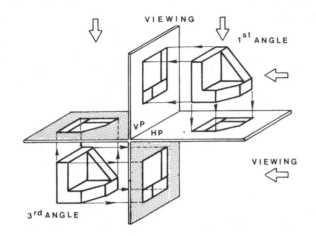

Fig. 5.11 Principle of orthographic projection

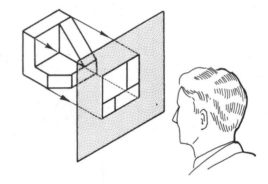

Fig. 5.13 Principle of third-angle projection

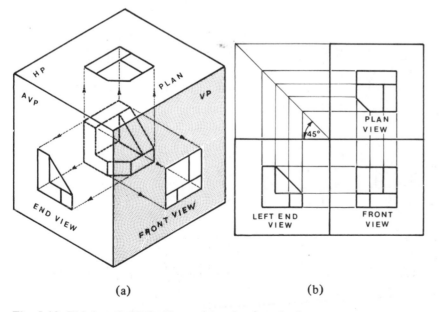

(a)   (b)

Fig. 5.12 Third-angle projection – 'glass-box' method

The symbol used to indicate third-angle projection is derived as for first-angle projection, but the views are positioned differently, as shown in fig. 5.14. The symbol shows a left end view and a front view of the circular taper in third-angle projection.

 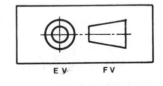

Fig. 5.14 Third-angle projection symbol

31

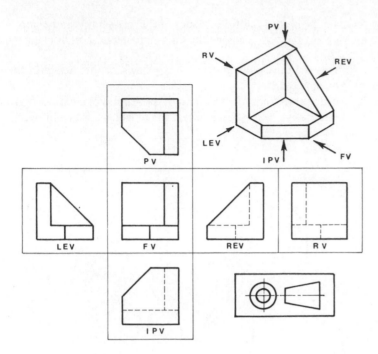

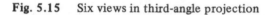

Fig. 5.15 Six views in third-angle projection

In common practice, only three of the six possible views (fig. 5.15) are used for a complete description of an object, as in fig. 5.16. Either of the two end views may be chosen.

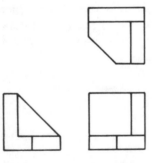

Fig. 5.16 Three views in third-angle projection

The advantage of third-angle projection is that the views drawn are positioned close to the surfaces or parts they represent. This is specially relevant with larger drawings.

5.4 Test questions

1. Figure 5.17 shows components A, B, C, D, E, and F in pictorial projection. The direction of viewing indicated by the arrows corresponds to the front views.

Select from the given orthographic views 1 to 18 the relevant front views (FV), end views (EV), and plan views (PV), and insert your answers in a table like the one provided.

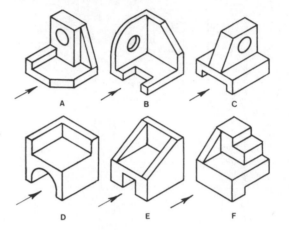

	FV	EV	PV
A	1	18	9
B			
C			
D			
E			
F			

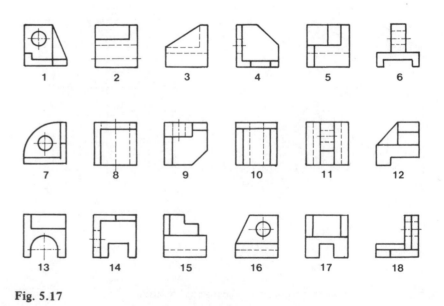

Fig. 5.17

32

2. a) Draw or sketch in third-angle projection the six views of the component shown in fig. 5.18(a). Name each view in full.
   b) Draw or sketch in third-angle projection the six views of the component shown in fig. 5.18(b).
   c) Draw or sketch in first-angle projection the six views of the component shown in fig. 5.18(c).

   Each construction square represents a 10 mm measurement.

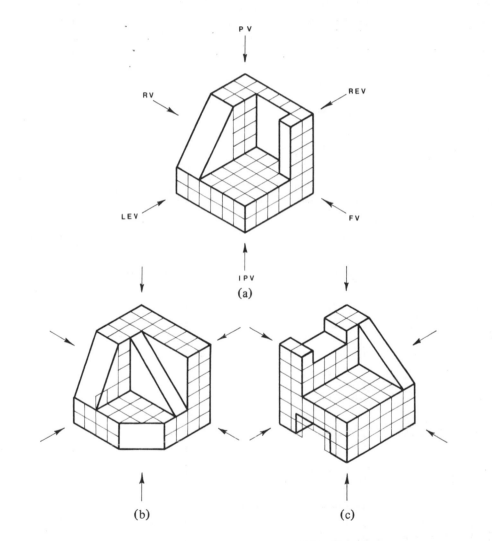

(a)

(b)   (c)

Fig. 5.18

3. Draw or sketch in third-angle projection the components shown in fig. 5.19. Select views to show most of the features as visible outlines. Each construction square represents a 10 mm measurement.

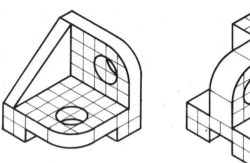

① ②

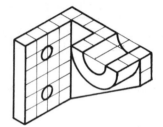
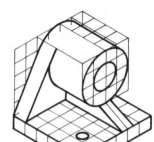

③ ④

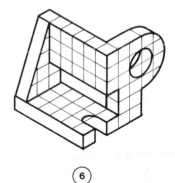
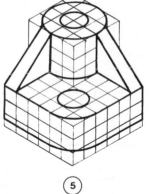

⑤ ⑥

Fig. 5.19

33

Fig. 5.20

4. In fig. 5.20, incomplete drawings 1 to 12 are in first-angle projection and 13 to 20 in third-angle. Complete each drawing by adding the missing lines. Tracing paper may be used.

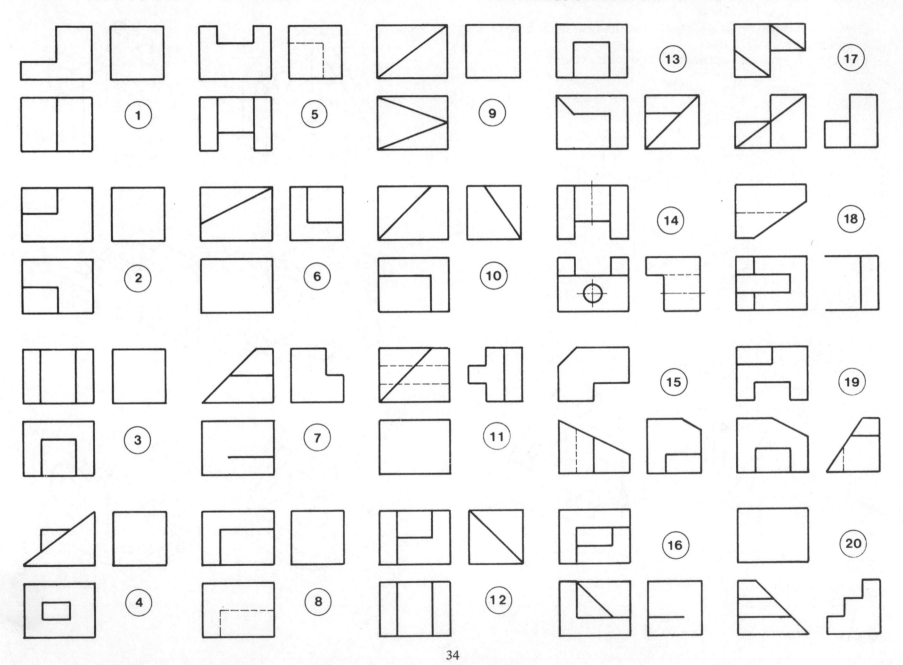

5. Complete the third view for each object shown in third-angle projection in
fig. 5.21. Squared or tracing paper may be used.

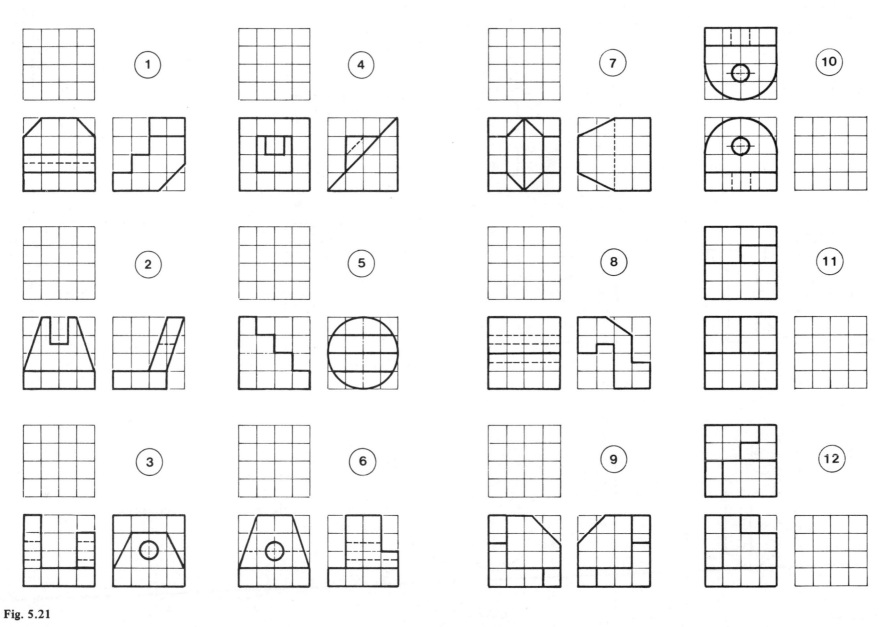

Fig. 5.21

35

6. Figure 5.22 shows a component in three different pictorial projections.
   a) Which is the correct perspective projection?
   b) Which is the correct isometric projection?
   c) Which is the correct oblique projection?

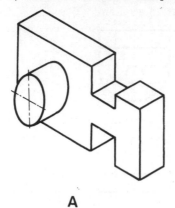

A

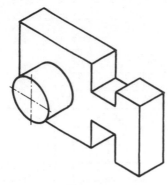

B

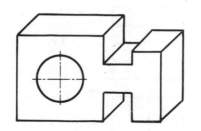

C

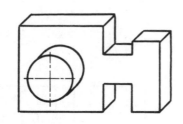

D

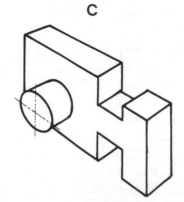

E

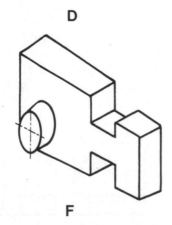

F

Fig. 5.22

7. Look at the drawings of a component shown in fig. 5.23.
   a) Which is in correct first-angle projection?
   b) Which is in correct third-angle projection?

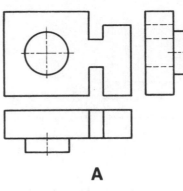

A

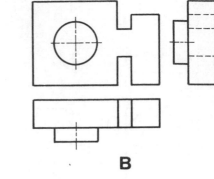

B

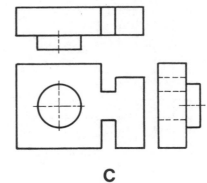

C

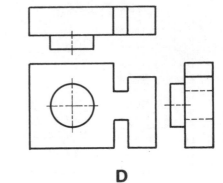

D

Fig. 5.23

8. In the figure below,
   a) Which view has to be deleted to obtain the correct symbol for first-angle projection?
   b) Which view has to be deleted to obtain the correct symbol for first-angle projection?

A   B

9. Select the correct end view — A, B, C, or D — for each object shown in fig. 5.24.

10. Sketch in first-angle projection objects 1 to 6 shown in fig. 4.22 on page 22.
11. Sketch in third-angle projection objects 7 to 12 shown in fig. 4.22 on page 22.
12. Draw in first-angle projection the objects shown in fig. 4.31 on page 25.
13. Draw in third-angle projection the objects shown in fig. 4.32 on page 25.

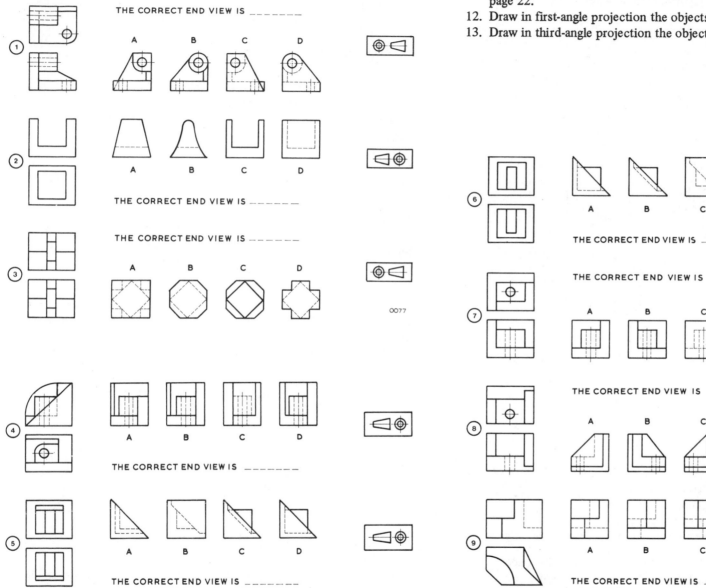

THE CORRECT END VIEW IS _____

A   B   C   D

THE CORRECT END VIEW IS _____

A   B   C   D

THE CORRECT END VIEW IS _____

A   B   C   D

0077

A   B   C   D

THE CORRECT END VIEW IS _____

A   B   C   D

THE CORRECT END VIEW IS _____

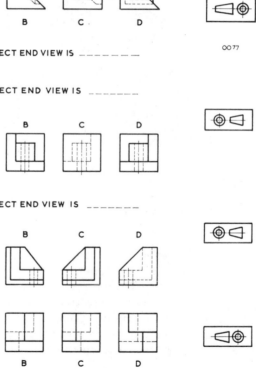

THE CORRECT END VIEW IS _____

A   B   C   D

0077

THE CORRECT END VIEW IS _____

A   B   C   D

THE CORRECT END VIEW IS _____

A   B   C   D

THE CORRECT END VIEW IS _____

A   B   C   D

THE CORRECT END VIEW IS _____

0077

Fig. 5.24

37

5.5 Sectional views

Quite often an outside view of an object does not adequately describe it, as no internal features are shown.

In order to show the internal features without excessive use of hidden-detail lines, the object is imagined to be cut along a plane called a *cutting plane*. The cut portion nearer to the observer is removed and the remaining part is shown as a sectional view.

The surfaces in section can be imagined to be cut along the cutting plane with an imaginary tool and imaginary cutting marks are represented by thin equidistant hatching lines as shown in fig. 5.25. Sometimes hatching may be omitted, if the clarity of drawing is not reduced by doing so.

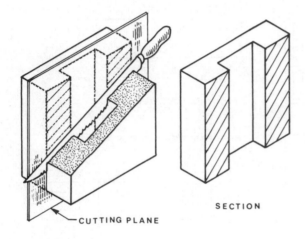

Fig. 5.25 Cutting plane

Sectional views are usually produced
a) to clarify details of the object,
b) to illustrate internal features clearly,
c) to reduce the number of hidden-detail lines,
d) to facilitate the dimensioning of internal features,
e) to show the shape of the cross-section,
f) to show clearly the relative positions of parts forming an assembly.

Cutting planes are represented on drawings by long thin chain lines thickened at each change of direction and at both ends. The direction of viewing is shown by arrows resting on thick lines at both ends, as in fig. 5.26.

Cutting planes should be designated by capital letters.

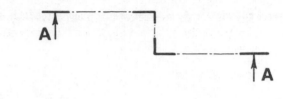

Fig. 5.26   Cutting plane AA

The surfaces shown in section are usually hatched at 45° or at some well defined angle which avoids clashing with visible outlines, as in fig. 5.27. Spacing between hatching lines should be equidistant and not less than 4 mm.

Fig. 5.27   Hatching

Adjacent components should be hatched in opposite directions. Hatching lines for additional adjacent parts can be offset or, alternatively, spacing between the lines may be increased or reduced, as shown for the internal part in fig. 5.28(a) and (b).

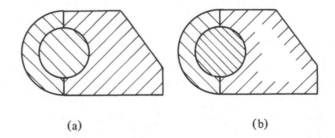

(a)   (b)

Fig. 5.28   Hatching of adjacent parts

Spacing between the hatching lines should be chosen in proportion to the size of the hatched section. In the case of large areas in section, the hatching may be limited to a zone following the contour of the hatched area, as shown in fig. 5.28(b).

38

When sections are shown side by side in parallel planes, as in fig. 5.29, hatching lines should be offset along the dividing chain lines between each section.

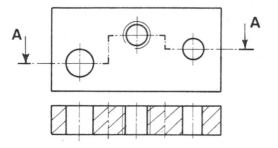

Fig. 5.29 Parallel cutting planes

Thin sections should be shown as single thick lines, leaving space between adjacent parts for clarity, as shown in fig. 5.30(a).

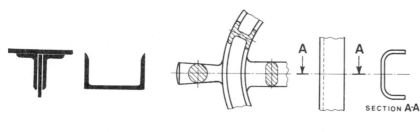

(a) Thin section (b) Revolved section (c) Removed section

Fig. 5.30

Cross-sections may be revolved in place, as in fig. 5.30(b). The outline is shown in continuous thin lines, and further identification is not necessary.

A cross-section may be removed as shown in fig. 5.30(c) — further identification is then necessary.

A symmetrical component may be shown in half section — in outside view to the centre line of symmetry and the remaining half in sectional view, as shown in fig. 5.31(a).

A local (part) section may be drawn to avoid the need for a separate sectional view. The local break is shown by a continuous thin irregular freehand line, as shown in fig. 5.31(b).

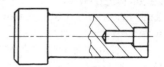

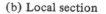

(a) Half section   (b) Local section

Fig. 5.31

There are several exceptions to the general rules of sectioning, as follows.

For clarity, the following features are not shown in section when cut longitudinally:

a) ribs and webs,
b) shafts, rods, and spindles,
c) bolts, nuts, and thin washers,
d) rivets, dowels, pins, and cotters,
e) spokes of wheels and similar parts.

If a shaft and web lie along the cutting plane, they are not sectioned, fig. 5.32(a). If a shaft and web lie across the cutting plane, as in fig. 5.32(b), then they are sectioned.

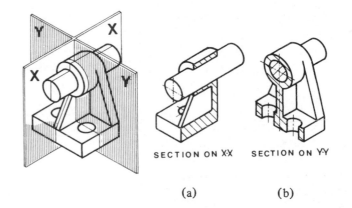

(a)   (b)

Fig. 5.32

The same sections on XX and YY are shown in orthographic projection in fig. 5.33.

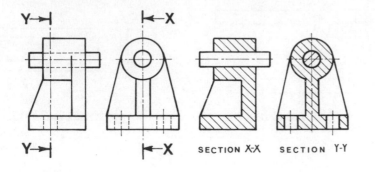

Fig. 5.33

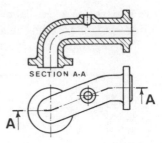

SECTION A-A

Fig. 5.35

In fig. 5.34(a), the cutting plane is revolved into the vertical position and is then projected to the sectional view.

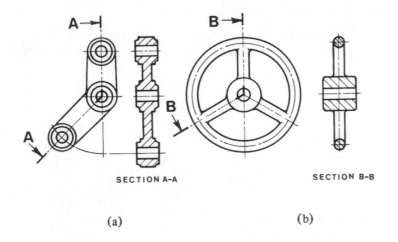

SECTION A-A

SECTION B-B

(a)   (b)

Fig. 5.34

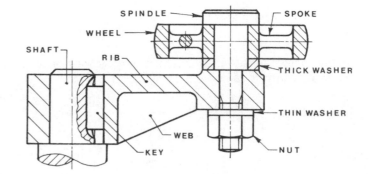

Fig. 5.36

Spokes of wheels and ribs are not sectioned longitudinally, as shown in fig. 5.34(b). The cutting plane is revolved into the vertical position and is then projected to the sectional view.

Figure 5.35 shows a section in three touching adjoining planes.

Figure 5.36 shows a section through an assembly which includes the features that are not usually sectioned.

When webs or ribs are cut along their length by a cutting plane, they are not sectioned, in order to avoid a false appearance of solidity. The webs and ribs are usually thin in comparison to the overall thickness of the main body.

If a cutting plane cuts across the webs or ribs, then they are shown in section in the usual way.

Nuts and bolts, thin washers, studs, screws, rivets, keys, pins, shafts, spindles, and spokes of wheels are more easily recognisable by their external features, so they are not shown in section if cut longitudinally (fig. 5.36).

5.6 Test questions

1. The components shown in fig. 5.37 are drawn in first- or third-angle projection. Sketch or redraw the two given views of each component and complete the third as a sectional view. Tracing paper may be used.

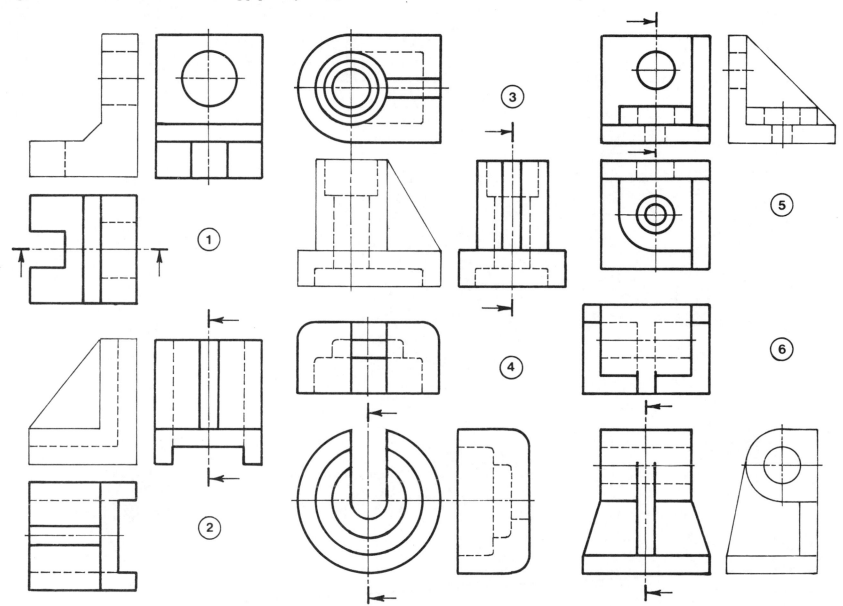

Fig. 5.37

2. For the components shown in fig. 5.38, complete all the end views in
   section as indicated. Tracing paper may be used.

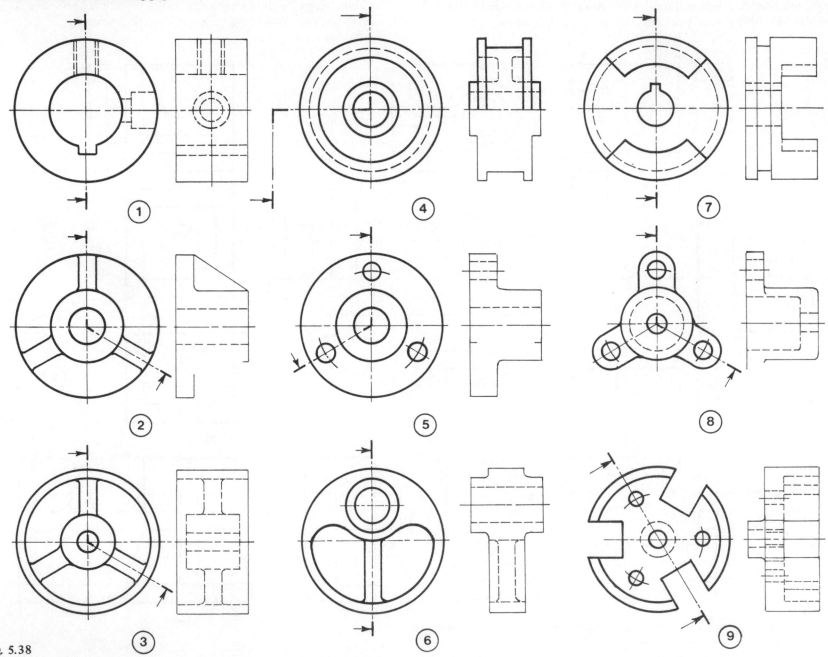

Fig. 5.38

42

5.7 Reading an engineering drawing

To read a drawing is to obtain a clear mental picture of what the person who prepared the drawing wishes to convey. Every engineer must be able to read and understand drawings.

In orthographic projection, at least two views are required for a full description of an object. Figure 5.39 shows that two given views do not necessarily describe an object completely, as several end views are possible.

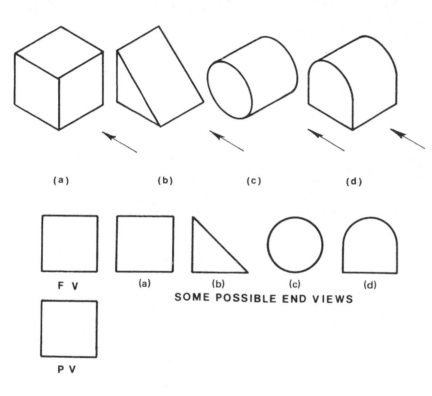

Fig. 5.39 Some possible end views for the given front and plan views

It is essential to acquire the ability to read drawings and to *visualise* the objects they represent. A drawing must be read patiently by referring systematically back and forth from one view to another. At the same time, the reader must imagine a three-dimensional object and not a two-dimensional flat projection.

A pictorial sketch usually helps to clarify the shape of a part that is difficult to visualise.

5.8 Test questions

1. Figure 5.40 shows two views in orthographic projection of each of six objects. Copy the given views using tracing paper, then complete the third views given and sketch each object in a pictorial projection in the space indicated.

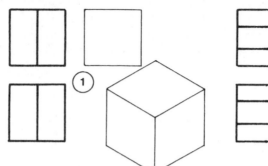
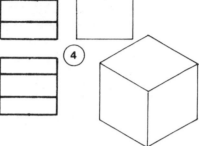
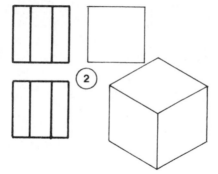
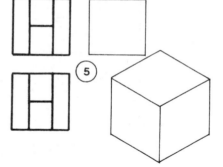
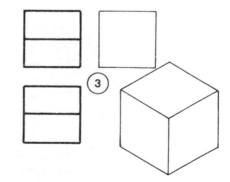
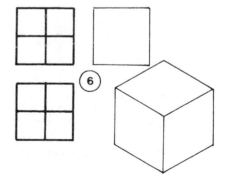

Fig. 5.40

43

2. For each of the objects 1 to 14 shown in orthographic projection in fig.
   5.41, copy the given views using tracing paper, then complete the
   unfinished third view and sketch the object in a pictorial projection.

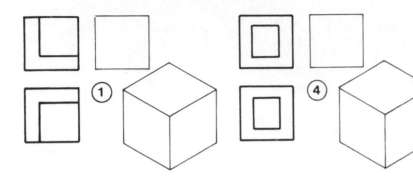

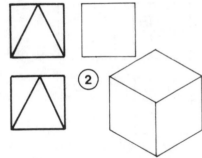

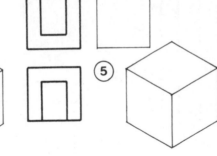

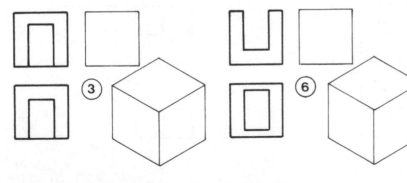

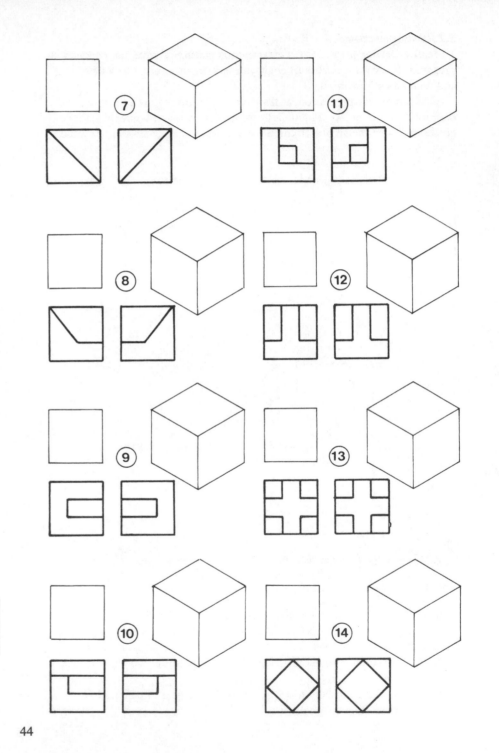

Fig. 5.41

5.9 Views on drawings

To ensure clear reading, due regard should be paid to the spacing of views on drawings — views and sections should not be overcrowded.

The number of views should be the minimum necessary to ensure that the drawing can be fully understood (see pages 1 and 2).

Views should be chosen so that as little information as possible has to be shown as hidden detail and to provide clearly and precisely the maximum possible information. In general, hidden detail should be used only where it is essential, and it should not be used for dimensioning purposes.

It is sometimes necessary to draw a partial view, which may be projected from an inclined feature of a component as shown in fig. 5.42.

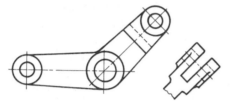

Fig. 5.42   Partial view

It is not necessary to draw symmetrical parts in full. As shown in fig. 5.43, a line of symmetry may be used — a thin chain line with two short thick parallel lines drawn at each end and at right angles to the symmetry line. The outline of the part is extended slightly beyond the symmetry line.

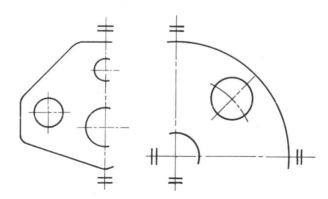

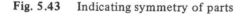

Fig. 5.43   Indicating symmetry of parts

Scale

Very large components are usually drawn to a reduced scale. To ensure clarity and precision, very small components are drawn larger than full size.

The drawing should be drawn in selected proportion to a uniform scale. This scale should be stated on the drawing as a ratio.

Scale multipliers and divisors of 2, 5, and 10 are recommended:

full size — 1:1

smaller than full size — 1:2, 1:5, 1:10, 1:20, 1:50, 1:100

larger than full size — 2:1, 5:1, 10:1, 20:1, 50:1, 100:1.

Some of the scales used are illustrated below.

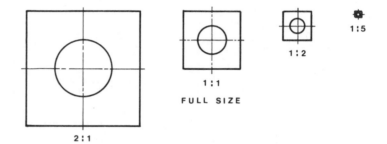

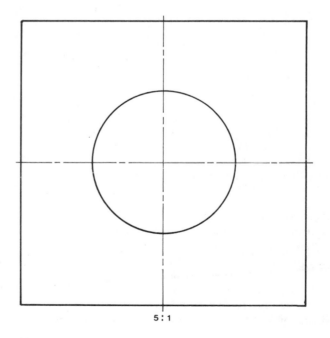

Drawing sequence (fig. 5.44)

A methodical approach will help to increase the speed of drawing and encourage accuracy and neatness.

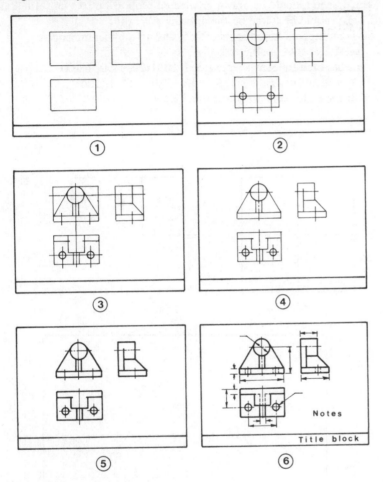

Fig. 5.44   Stages in drawing preparation

Drawing faintly,
1. space the overall sizes of the views on the drawing paper;
2. insert centre lines and draw the circles and arcs;
3. draw the required views;
4. check everything, correct mistakes, and remove all unwanted lines etc.

With increased pencil pressure,
5. complete the final lining in — circles and arcs first;
6. insert all required dimensions, notes, and sections if required.

To draw a line parallel to a given straight line AB at a distance R from it (fig 5.45)
1. Set compasses at R and draw two arcs with centres on AB.
2. Draw a tangential line CD to touch both arcs.
   CD is the required line.

Fig. 5.45

To draw a fillet, radius R, tangential to lines AB and CD (fig. 5.46)
1. Draw line GE parallel to AB and distance R from it.
2. Draw line EF parallel to CD and distance R from it.
3. The intersection of GE and EF gives the centre for the required fillet radius, which is drawn between two centre lines normal to AB and CD as shown.

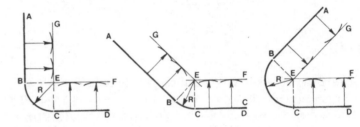

Fig. 5.46

To divide a line AB of any length into a number of equal parts (fig. 5.47)
1. Draw line AC to a convenient length at any angle to AB and divide AC into the required number of equal parts (say 6).
2. Join the last point, 6, on AC to the end of AB and draw lines parallel to 6B through points 1 to 6 by sliding a set square along a straight edge. These lines divide AB into the required equal parts.

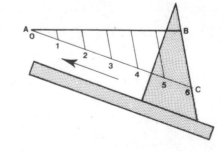

Fig. 5.47

5.10 Test questions

1. Draw or sketch twice full size (2:1) in first-angle projection the following views of the bracket shown in fig. 5.48(a):

   a) a sectional front view on AA,
   b) a sectional plan view on BB,
   c) an end view.

   Ensure the correct positioning of all views.
   Each construction square represents a 5 mm measurement.

2. Draw or sketch half full size (1:2) in third-angle projection the following views of the bracket shown in fig. 5.48(b):

   a) a sectional front view on AA,
   b) an end view,
   c) a plan.

   Ensure the correct positioning of all views.
   Each construction square represents a 20 mm measurement.

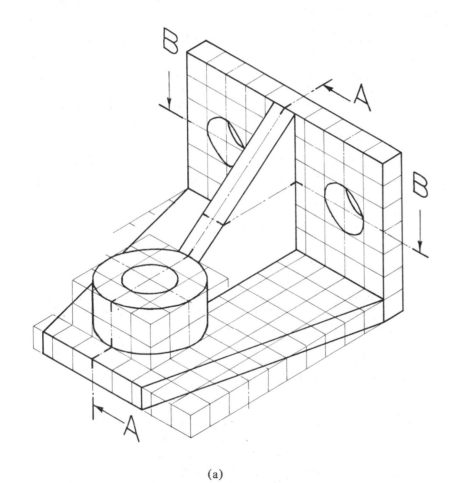

(a)

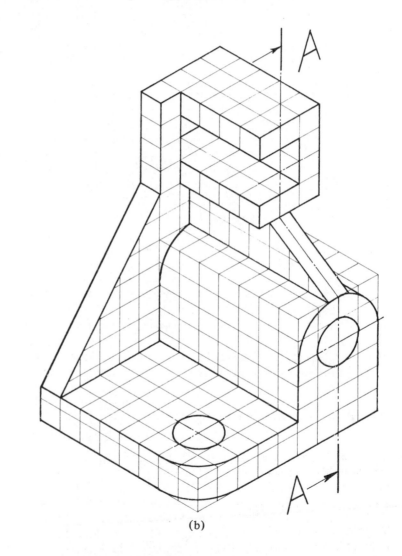

(b)

Fig. 5.48

5.11 Points, lines, and plane surfaces in space

Figure 5.49(a) shows two points A and B suspended in the space between three principal planes drawn in first-angle projection.

The views of each point are projected to the principal planes and are indicated by the lower-case letters a and b, as if they were shadows of the points A and B.

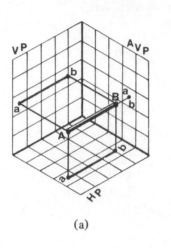

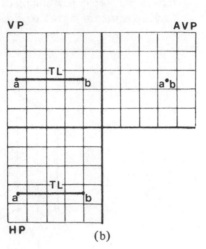

(a)

(b)

Fig. 5.49

Figure 5.49(b) shows all three planes unfolded, and three views of the points A and B are shown simultaneously.

Now consider a straight line AB. This is projected as the lines ab on VP and HP but only as a point on AVP.

As the line AB is parallel to the vertical and horizontal principal planes (VP and HP), the views in those planes represent the true length (TL) of the line.

The rule When a line is suspended in space, the view of the line projected on to any plane parallel to the line will represent the true length of the line.

Note: For clarity the line AB has been given thickness, as though it were a thin rod.

Figure 5.50 shows a line AB parallel to the horizontal plane (HP). The plan view on the horizontal plane is the true length (TL) of the line. The two remaining views show foreshortened lines and not true lengths.

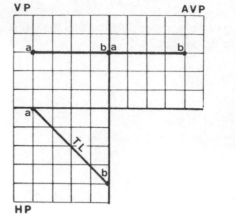

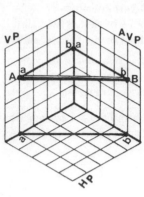

Fig. 5.50

Figure 5.51 shows a line AB parallel to the auxiliary vertical plane (AVP). The end view on AVP is the true length of the line.

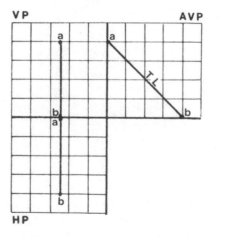

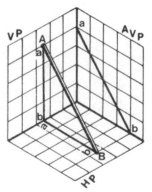

Fig. 5.51

48

Figure 5.52 shows a line AB which is not parallel to any of the principal planes, therefore none of the views on these planes represents the *true length* of the line.

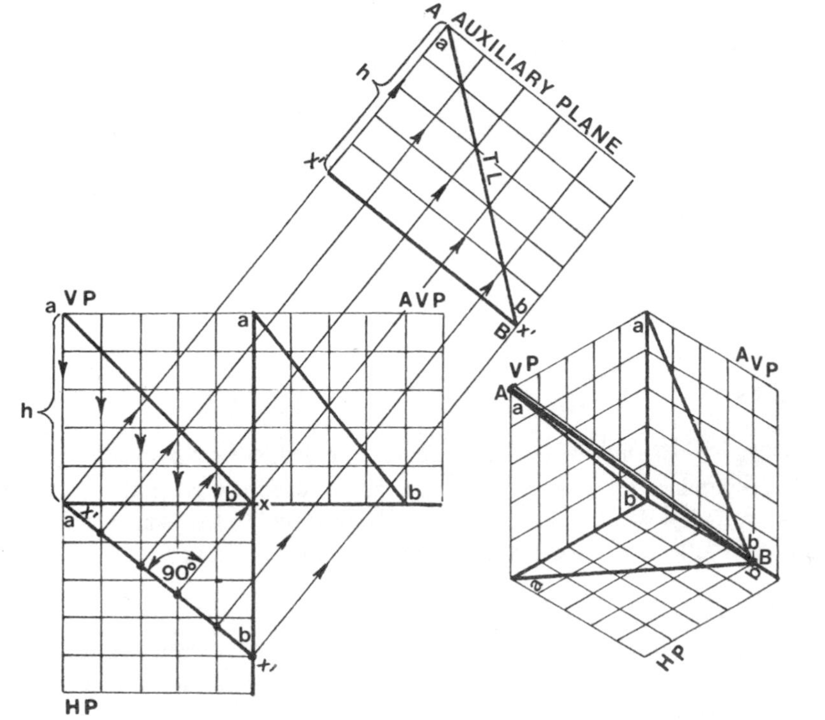

Figure 5.53 shows two lines AB and AC which are not parallel to any of the principal planes.

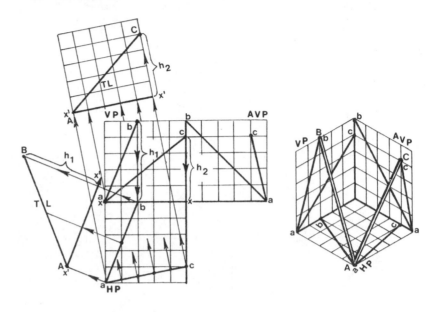

Fig. 5.52

Fig. 5.53

If the true length is to be shown, a new plane parallel to the line AB has to be introduced. This plane is called an *auxiliary plane*.

To draw the true length

1. Draw an auxiliary plane parallel to the line ab in the plan view (on HP), projected at right angles to that line.
2. Transfer all vertical distances *h* measured between the reference line xx and the line ab in the front view (on VP) to the auxiliary plane by reflecting at 90° from the line ab in the plan view (on HP) and measuring from x′x′.
3. The auxiliary view on AVP will be the true length of the line AB.

To find the true lengths of the lines AB and AC, the required distances are transferred from the front view (on VP) to the plan view (on HP) and then reflected to the respective auxiliary views.

It is not necessary to draw a complete auxiliary plane in each case — it may be omitted as shown for the line AB, where only the distances from the reference line x′x′ are used.

Figure 5.54 shows a triangular shape ABC. The true lengths are obtained by transferring all required vertical distances from the front view (on VP) to the auxiliary views.

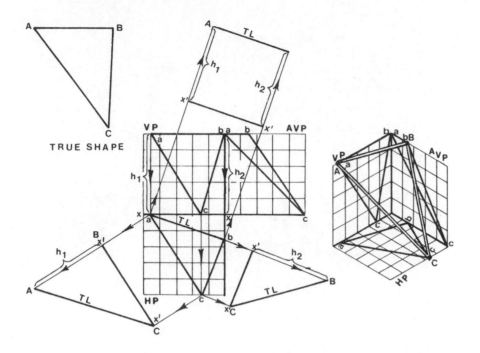

Fig. 5.54

Using the true lengths obtained in the auxiliary views, the true shape of ABC is constructed as follows.

1. Draw the true length AB.
2. From centre A with true length AC as radius, strike an arc.
3. From centre B with true length BC as radius, strike an arc.
4. The intersection of these arcs is the required point C.
5. Join AC and BC.

Figure 5.55(a) shows the line AB drawn in third-angle projection inside a 'glass box'. All the projected views of the line AB are viewed through the 'transparent' principal planes.

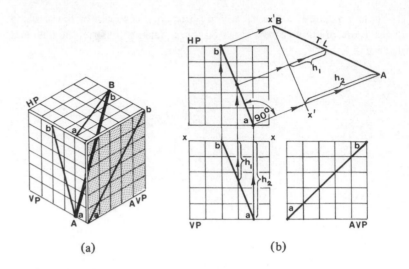

(a)   (b)

Fig. 5.55

Figure 5.55(b) shows the three planes unfolded, where the true length (TL) is obtained using the same method as for first-angle projection: from the front view (on VP), vertical distances are transferred to the plan view (on HP) and then reflected to the auxiliary view and measured from x'x'.

5.12 Test questions
Complete all the views shown in fig. 5.56(a) and at 1, 2, 3, and 4 in fig. 5.56(b) and determine the true length of each line AB. Also draw the true shapes for figures 2, 3, and 4 in fig. 5.56(b).

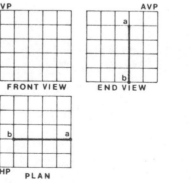
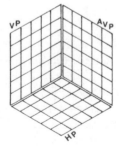

Fig. 5.56 (a)

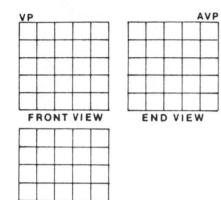

VP AVP

FRONT VIEW END VIEW

HP PLAN

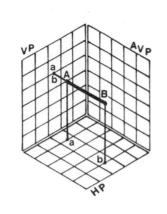

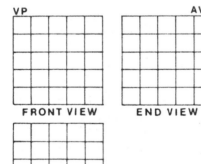

①

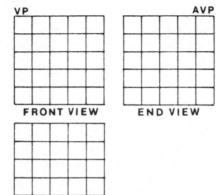

VP AVP

FRONT VIEW END VIEW

HP PLAN

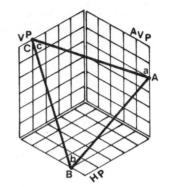

②

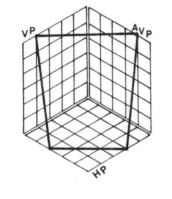

VP AVP

FRONT VIEW END VIEW

HP PLAN

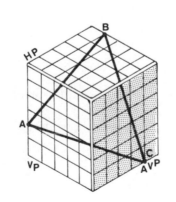

③

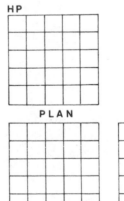

④

Fig. 5.56 (b)

51

5.13 Auxiliary views

Occasionally a component has surfaces which are not parallel with any of the principal planes of orthographic projection and which therefore cannot be clearly defined or dimensioned. To draw the true shapes of those surfaces, additional views are required showing the surfaces as they appear when looking directly at them. These views are called *auxiliary views*.

Figure 5.57 shows an object suspended inside a third-angle-projection 'glass box' which consists of three principal planes and two auxiliary planes.

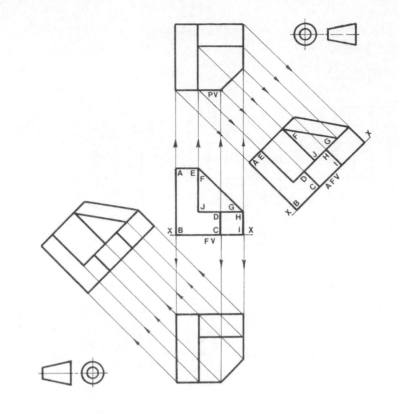

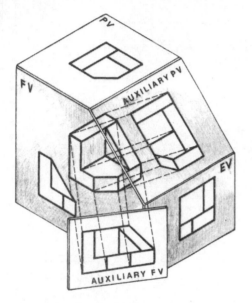

Fig. 5.57 Auxiliary planes

The auxiliary horizontal plane shows the auxiliary plan view (APV) and the auxiliary vertical plane shows the auxiliary front view (AFV).

Figure 5.58 shows how the auxiliary front view is obtained in third- and first-angle projections.

Fig. 5.58 Auxiliary front view

To draw the auxiliary front view (fig. 5.58)

1. Draw the front view (FV) and plan view (PV).
2. Draw the chosen reference line XX in the front view and similarly in the auxiliary front view at right angles to the direction of viewing.
3. Transfer all required vertical distances measured from XX in the front view along projectors to the relevant points in the plan view and reflect them to the auxiliary view. Measure the same distances from the reference line XX in the auxiliary view.
4. Join the points so obtained to complete the required auxiliary front view (AFV).

Figure 5.59 shows how the auxiliary plan view is obtained in third- and first-angle projections.

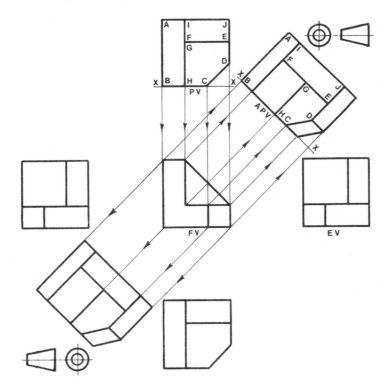

Fig. 5.59 Auxiliary plan view

To draw the auxiliary plan view (see fig. 5.59)

1. Draw the front and plan views.
2. Draw the chosen reference line XX in the plan view and similarly in the auxiliary plan view at right angles to the direction of viewing.
3. Transfer all required vertical distances measured from XX in the plan view along projectors to the relevant points in the front view and reflect them to the auxiliary view. Measure the same distances from the reference line XX in the auxiliary view.
4. Join the points so obtained to complete the required auxiliary plan view (APV).

5.14 Test questions

1. Draw an auxiliary view of each object shown in fig. 5.60, looking in the direction of arrow Y. To clarify the drawing of auxiliary views, some of the points are indicated with capital letters. The letters with dashes refer only to points on the base of the objects.

Note that, in order to draw a surface or edge in the auxiliary view, all projectors must be reflected from that particular surface or edge in the front or plan view before the measurements can be taken, as shown in figure 4.

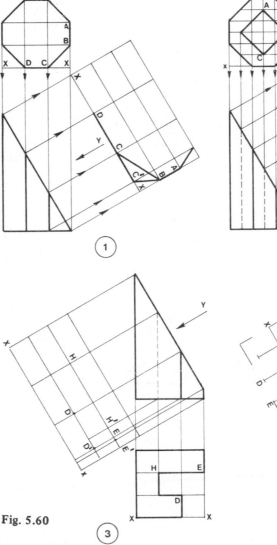

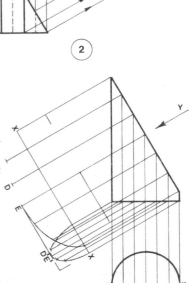

Fig. 5.60

2. Draw an auxiliary view of each component in fig. 5.61, looking in the direction shown. Tracing paper may be used.

3. Draw an auxiliary view of each component shown in fig. 5.62, looking in the direction of the arrow normal to the inclined surface. Each construction square represents a 10 mm measurement.

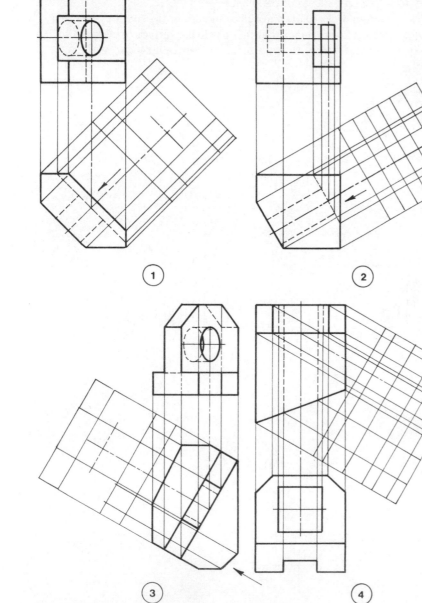

Fig. 5.61

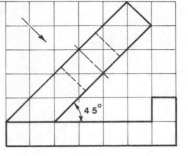
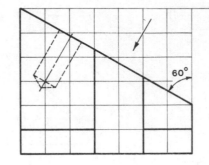

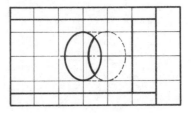
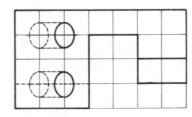

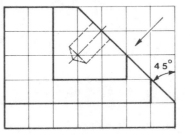
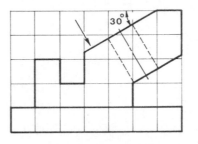

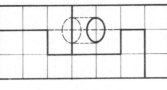
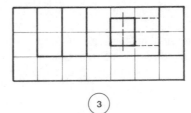

Fig. 5.62

4. Four separate drawings — A, B, C, and D — of a solid cylinder are shown in fig. 5.63. Each drawing is in orthographic projection and consists only of a front view, an auxiliary view, and a projection symbol. In which drawings is the projection symbol used correctly?

5. Figure 5.64 shows four separate drawings — A, B, C, and D — of a five-sided bar, machined as shown. Each drawing is in orthographic projection and consists only of a front view, an auxiliary view projected from the front view, and a projection symbol. In which drawings is the projection symbol used correctly?

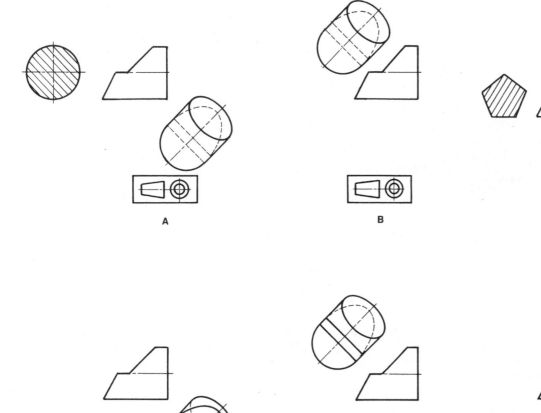

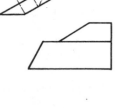

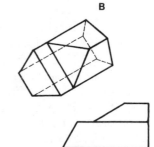

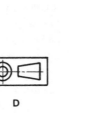

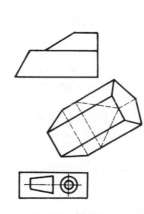

Fig. 5.63

Fig. 5.64

6 Dimensioning

An engineering drawing conveys information in two ways:

a) by a pictorial or orthographic *view* of the object,
b) by *instructions* in the form of given sizes or dimensions and notes specifying the manufacturing processes and materials.

Dimensions can be considered to be of two types:

i) those which define the size and shape of an object or feature — called *size dimensions*;
ii) those which specify the relative positions of various features — called *location dimensions*.

In addition, dimensions can be put into three groups relative to the function of a product: functional dimensions, non-functional dimensions, and auxiliary dimensions.

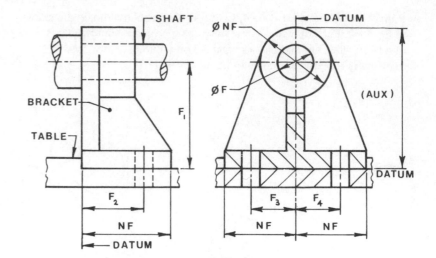

Fig. 6.1 Functional, non-functional, and auxiliary dimensions

Functional dimensions

These dimensions directly affect the function or working of a product and may be of the size or location types.

Functional dimensions should be based on the function of the component and they can also show the method of locating the component in its appropriate assembly, thus ensuring its correct working within the whole product.

A *datum* is a reference line on the drawing from which a component is dimensioned. In practice, a datum is any functional surface or axis used for manufacture, inspection, location, or assembly purposes. To ensure the required accuracy of measurement, the datum surfaces are machined to the required degree of finish.

The main function of the components in the assembly shown in fig. 6.1 is to support a shaft. Hence the functional location dimension, F_1, is between the hole centre and the datum mating face.

The fixing holes must be positioned in relation to the shaft and the shoulder on the mating face of the table. The functional dimension F_2 is used for this purpose.

The same dimension F_2 will apply to the bracket and the table. Also, the centre lines of the holes must be positioned relative to the vertical datum using the functional dimensions F_3 and F_4.

Non-functional dimensions

Non-functional dimensions, NF in fig. 6.1, are those dimensions which are used for production and inspection purposes but which do not directly affect the function or working of a product.

Auxiliary dimensions

Auxiliary dimensions, (AUX) in fig. 6.1, are given for information only. They are not used for production or inspection purposes, they should not be toleranced, and they should always be inserted in parentheses (brackets).

Auxiliary dimensions are *redundant* dimensions which provide useful information but do not govern acceptance of the product.

6.4 Principles of dimensioning

Dimensions are normally expressed in millimetres. The decimal point should be bold and placed on the base line of the numbers. Dimensions of less than unity should be preceded by zero, e.g. 0.6 mm.

Every drawing should include only as many dimensions as are necessary to define the component.

Each dimension should appear only once — it should not be repeated on other views.

It should not be necessary to calculate any dimension from other dimensions.

In an industrial situation, it should not be necessary for the drawing to be scaled to deduce a dimension. In a college, for exercise purposes only, drawings may be scaled or traced to save valuable time.

Dimensions relative to a particular feature should be placed in one view, rather than spread over several views.

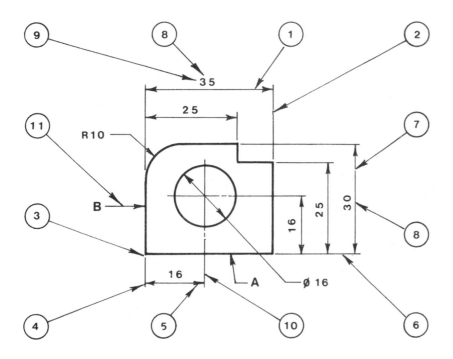

Fig. 6.2

In fig. 6.2, A and B are the reference edges which are used as datums for dimensioning.

The following are explanatory notes referring to fig. 6.2.

1. Dimension lines are thin continuous lines and for preference should be placed outside the component.
2. Projection lines are thin continuous lines projected from outlines. The crossing of projection and dimension lines with other lines should be kept to a minimum.
3. A small gap should be left between the outline and the start of a projection line.

4. Extension lines should continue the projection lines slightly beyond the dimension line.
5. Dimension lines should end in arrowheads and must touch the extension lines. Arrowheads should be about 3 mm long, slender, and closed at thick ends.
6. Dimension lines should be well spaced, equidistant, and placed outside the outlines of the component.
7. Smaller dimensions should be placed nearest to the outlines of the component and larger dimensions outside smaller dimensions.
8. Numerals should be placed so that they may be read from the bottom or from the right-hand side of the drawing.
9. Numerals or letters should preferably be placed centrally above and clear of their dimension lines.
10. A centre line, outline, or projection line should never be used as a dimension line, but a centre line may be used as a projection line.

Leaders

Leaders are thin continuous lines indicating the outlines or surfaces to which relevant dimensions or notes apply.

Leaders end in arrowheads when touching and stopping on a line but in dots when crossing the line, as in fig. 6.3(a).

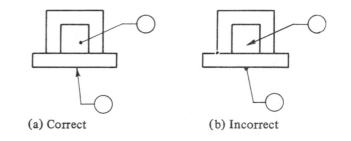

(a) Correct   (b) Incorrect

Fig. 6.3

Leaders touching lines should be nearly normal (at right angles) to those lines, as shown in fig. 6.4(a).

(a) Correct   (b) Incorrect

Fig. 6.4

Dimensioning common features

When notes or dimensions refer to *repeated features*, a long or intersecting leader should not be used. The dimensions should be repeated as in fig. 6.5(a) or letter symbols should be used as in fig. 6.6(a).

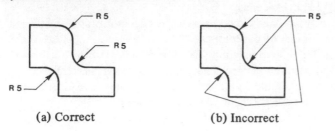

(a) Correct   (b) Incorrect

Fig. 6.5 Dimensioning repeated features

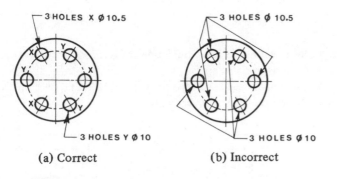

(a) Correct   (b) Incorrect

Fig. 6.6 Dimensioning repeated features

All *radius* dimension lines should pass through or be in line with the centre of the arcs, and they should have only one arrowhead, touching that arc.

Figure 6.7(a) shows how radii of arcs are dimensioned with centres located and not located.

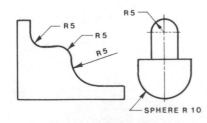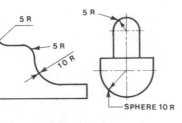

(a) Correct   (b) Incorrect

Fig. 6.7 Dimensioning radii

Dimensioning of *angular positions* of holes on a pitch circle is shown in fig. 6.8(a).

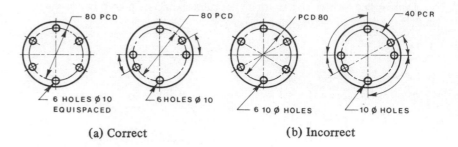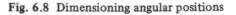

(a) Correct   (b) Incorrect

Fig. 6.8 Dimensioning angular positions

Placing of *angular dimensions* is shown in fig. 6.9(a). The shaded area is used only to highlight the position of the corresponding numerals.

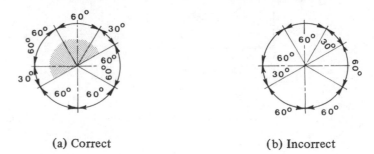

(a) Correct   (b) Incorrect

Fig. 6.9 Placing angular dimensions

Chamfers of 45° and chamfers at angles other than 45° are dimensioned as shown in fig. 6.10(a).

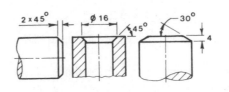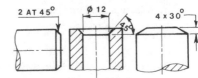

(a) Correct   (b) Incorrect

Fig. 6.10 Dimensioning chamfers

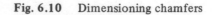

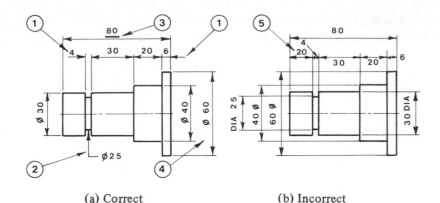

(a) Correct   (b) Incorrect

Fig. 6.11

Figure 6.11(a) shows the correct dimensioning of *diameters* and other features of a component:

1. When dimensioning small features, the numerals should be placed centrally or above the extension of one of the arrowheads. The narrow space between arrowheads may or may not include a line.
2. For clarity, leaders may be used for dimensioning diameters.
3. Dimensions not drawn to scale should be underlined.
4. Symbol ∅ should be in front of a dimension giving the diameter of a circle or cylinder. The small circle of the symbol should be as large as the following numerals, and the sloping line passing through it should be at 60° to the horizontal.
5. Where an overall dimension is shown, one of the intermediate dimensions should be omitted as *redundant*.

Figure 6.12(i)(a) shows four methods of dimensioning *circles*. When a leader touches a circle, it should be in line with the centre of that circle.

The diameter of a *spherical surface* should be dimensioned as shown in fig. 6.12(ii)(a).

Some methods of dimensioning countersinks and counterbores and of specifying tapered features on a drawing are shown in fig. 6.13(a).

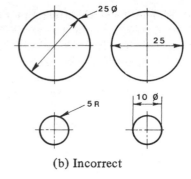
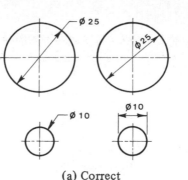

(a) Correct   (b) Incorrect

Fig. 6.12 (i) Dimensioning circles

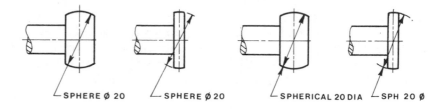

(a) Correct   (b) Incorrect

Fig. 6.12 (ii) Dimensioning spherical surfaces

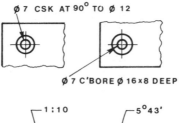
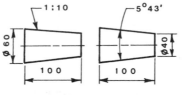
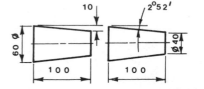

(a) Correct   (b) Incorrect

Fig. 6.13 Dimensioning countersinks, counterbores, and tapered features

59

6.5 Dimensioning for different purposes

Every drawing must include as many dimensions as are necessary for the complete definition of the component, describing the size, shape, and positioning of all relevant features.

Sometimes a component requires several drawings for different stages of manufacture, with corresponding dimensions.

6.6 Dimensioning for primary production

For primary production — which may include casting, forging, or fabrication — the drawing should have all dimensions necessary for shaping the object at that stage of production.

Sand casting

In sand casting, a cavity having the shape of the required component is formed in a box of sand by a wooden mould called a pattern. Molten metal is poured into the cavity and is allowed to cool. The casting is then removed and is ready to be machined. In sand casting, depending on the brittleness, castability, and strength of the materials, the casting usually has strengthening ribs or webs and blending fillets. When dimensioning, fig. 6.14, an allowance for shrinkage and warping has usually to be taken into account.

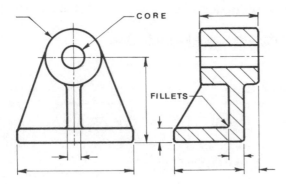

Fig. 6.14   Dimensioning for sand casting

Die casting

In die casting, the molten metal is forced into a mould under pressure, making it possible to get the surfaces more true and smooth and the dimensions more accurate than in sand casting, as there is less shrinkage and warping.

Drop forging

In drop forging, the metal is heated and then hammered into the desired shape, using a die.

In dimensioning, fig. 6.15, all corners of the component are rounded to allow correct metal formation, and the sides are usually tapered to facilitate the removal of the component from the die after forging.

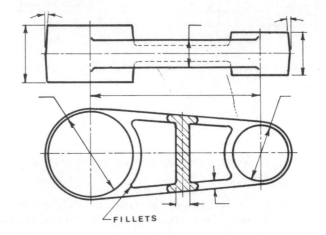

Fig. 6.15   Dimensioning for drop forging

Fabrication

In fabrication, a welded construction is used for small quantities of components which usually cannot be cast because of strength requirements and cannot readily be forged because of their shape or size.

Fabricated components are built of several different pieces of steel cut to shape and then welded together. For dimensioning, fig. 6.16, the shapes and relative positions of all separate pieces have to be taken into consideration, and all welding instructions must be clearly stated (see pages 82–3).

6.7 Dimensioning for secondary production

For the purposes of secondary production, which may include any machining processes, the drawing should have only those dimensions necessary for machining the required surfaces, as in fig. 6.17. (See also section 6.9, page 67).

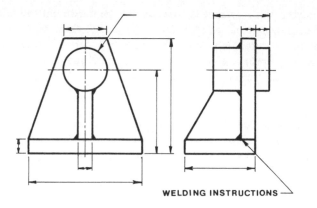

Fig. 6.16   Dimensioning for fabrication

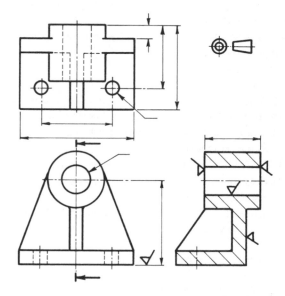

Fig. 6.17   Dimensioning for machining

Inspection

For the purposes of inspection, which is to ensure that components are within the dimensional limits laid down, the drawing will include all dimensions necessary to indicate the maximum and minimum limits of the machined sizes and the roughness of surfaces if required.

Rapid and economic inspection of limits is carried out by means of different types of gauge. The principle of limit gauging is that work is acceptable if the 'GO' section of a limit gap gauge passes over the work and the 'NO GO' section does not pass, fig. 6.18.

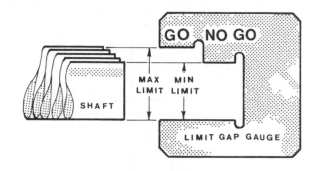

Fig. 6.18   Principle of limit gauging

Functional dimensioning

For functional purposes, the drawing should include all functional dimensions that have a direct bearing on the function and working of the components or their relative location, as in fig. 6.19.

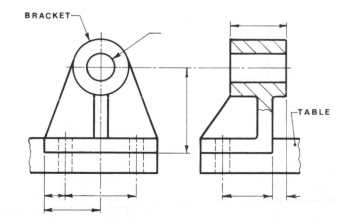

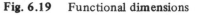

Fig. 6.19   Functional dimensions

Assembly

For assembly purposes the drawing may include overall and fitting dimensions. Sometimes the functional dimensions are included in this type of drawing, as shown in fig. 6.20.

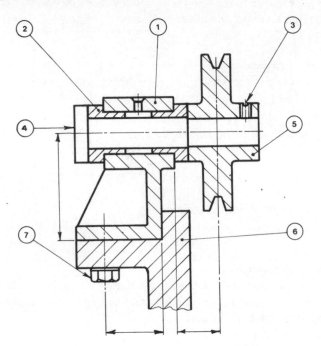

1	BRACKET	1
2	BUSH	2
3	GRUB SCREW	1
4	SHAFT	1
5	PULLEY	1
6	TABLE	1
7	BOLT AND NUT	2

Fig. 6.20   Assembly with parts list

6.8 Limits and fits

It is impossible to manufacture a component to the exact design size, this size being a numerical value of length. To overcome this difficulty, a tolerance is permitted or 'tolerated' which is the amount of deviation from the given basic design size, or the margin of error, allowable to accommodate reasonable inaccuracy in manufacturing.

On a drawing, this tolerance is indicated by the maximum and minimum permitted sizes, which are called *limits of size*, as shown in fig. 6.21. Tolerances should be stated only where the required accuracy is essential.

There are three principle considerations in establishing reasonable tolerances: function, interchangeability, and cost.

Function

The tolerances must be consistent with the design and function or working of the component. If a shaft is designed to rotate in a bearing, the tolerances must ensure that the shaft functions properly and is never larger than the hole of the bearing.

Interchangeability

Mass production often depends on producing certain parts on different machines to predetermined tolerances to ensure that all assembled parts will always fit together to form properly operating units. This can be controlled and communicated to the workshops only through the detail working drawings, and interchangeability can be ensured through inspection utilising an efficient system of gauging.

Cost

The expense of manufacture varies directly with the accuracy required; hence tolerances should be as large as the design will permit without affecting the function of the component. Large tolerances introduce savings in labour, use of tools, and running of machines.

In general, limits of size should be applied only if the specified accuracy is essential for the efficient functioning of the components or to facilitate their interchangeability.

Figure 6.21 shows the elements of an interchangeable system where two parts fit together — an external part (a shaft) and an internal feature of a part (a hole).

Basic size is usually a theoretical design size to which all limits of size are referred.

It is the same for both members of the fit.

Deviation is the algebraic difference between an actual size, obtainable by measurement, and the corresponding basic size.

Upper deviation is the algebraic difference between the maximum limit of size and the corresponding basic size.

Lower deviation is the algebraic difference between the minimum limit of size and the corresponding basic size.

Zero line is the line of zero deviation and represents the basic size.

Maximum limit of size is the maximum size permitted for a feature.

Minimum limit of size is the minimum size permitted for a feature.

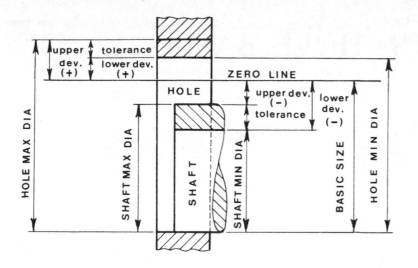

Fig. 6.21

A designer must ensure that an assembly of mating components will function correctly, thus all parts must fit together in the required manner.

A particular fit will depend solely on the prescribed maximum and minimum limits of size of the two separate components which are to be assembled. Engineering fits can be divided into three main types: clearance fits, interference fits, and transition fits.

Clearance fit

This is a fit which provides a clearance; hence the shaft is always smaller than the hole into which it fits, as in fig. 6.22(a). *Clearance* is the positive difference between the sizes of the hole and the shaft.

Typical applications of the clearance fit are on rotating shafts, loose pulleys, fast pulleys, bearings, cross-head slides, etc.

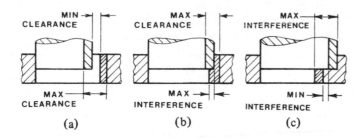

Fig. 6.22

Interference fit

This is a fit which always provides an interference; hence the shaft is always bigger than the hole into which it fits, as in fig. 6.22(c). *Interference* is the negative difference between the sizes of the hole and the shaft.

Typical applications of the interference fit are on pressed-in bushes or sleeves, crank pins, shrunk-on couplings, iron tyres, railway wheels shrunk on to axles, etc.

Transition fit

This is a fit which may provide either a clearance or an interference; hence the shaft may be bigger, smaller, or the same size as the hole into which it fits, as in fig. 6.22(b).

Typical applications of the transition fit are on bushes, spigots, fasteners, pins, keys, stationary parts for location purposes, etc.

A system of limits and fits may be on a hole basis or a shaft basis.

Hole-basis system

This is a system of fits in which the basic diameter of the hole is constant while the shaft size varies with different types of fit, see fig. 6.23(a). The minimum limit of hole size is the basic size.

The hole basis is more economical than the shaft basis as only one size of drill or reamer need be used to produce different fits, the shafts being turned and ground to the required sizes, thus making manufacture and measurement much easier.

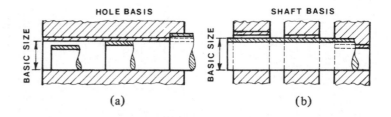

Fig. 6.23

Shaft-basis system

This is a system of fits in which the hole size is varied to produce the required type of fit, with the basic diameter of the shaft being constant. The maximum limit of the shaft is the basic size, see fig. 6.23(b).

This system tends to be less economical, as a series of drills is required. It is usually adopted where a single driving shaft accommodates a number of pulleys, bearings, collars, couplings, etc.

BRITISH STANDARD
SELECTED ISO FITS—HOLE BASIS

Data Sheet 4500A · Issue 1. February 1970

Diagram to scale for 25 mm. diameter

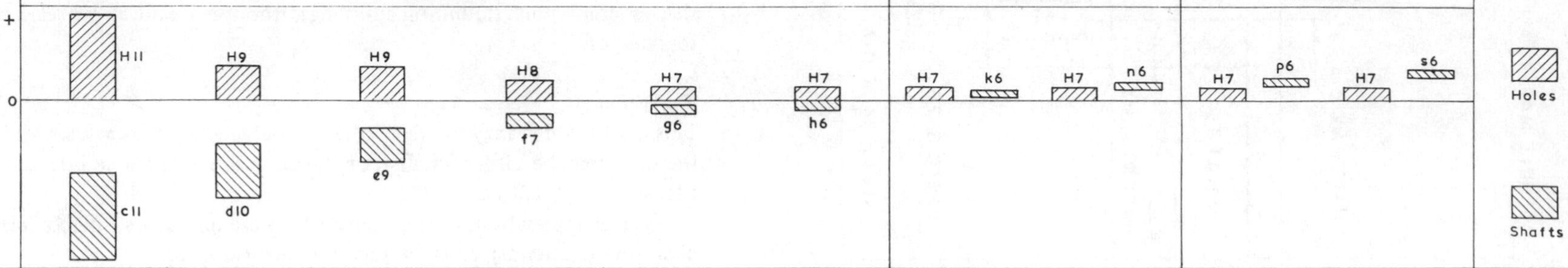

Clearance fits: H11, H9, H9, H8, H7, H7 (holes); c11, d10, e9, f7, g6, h6 (shafts)
Transition fits: H7, k6; H7, n6
Interference fits: H7, p6; H7, s6

Holes / Shafts

Over (mm)	To (mm)	H11	c11	H9	d10	H9	e9	H8	f7	H7	g6	H7	h6	H7	k6	H7	n6	H7	p6	H7	s6	Over (mm)	To (mm)
—	3	+60 / 0	−60 / −120	+25 / 0	−20 / −60	+25 / 0	−14 / −39	+14 / 0	−6 / −16	+10 / 0	−2 / −8	+10 / 0	−6 / 0	+10 / 0	+6 / +0	+10 / 0	+10 / +4	+10 / 0	+12 / +6	+10 / 0	+20 / +14	—	3
3	6	+75 / 0	−70 / −145	+30 / 0	−30 / −78	+30 / 0	−20 / −50	+18 / 0	−10 / −22	+12 / 0	−4 / −12	+12 / 0	−8 / 0	+12 / 0	+9 / +1	+12 / 0	+16 / +8	+12 / 0	+20 / +12	+12 / 0	+27 / +19	3	6
6	10	+90 / 0	−80 / −170	+36 / 0	−40 / −98	+36 / 0	−25 / −61	+22 / 0	−13 / −28	+15 / 0	−5 / −14	+15 / 0	−9 / 0	+15 / 0	+10 / +1	+15 / 0	+19 / +10	+15 / 0	+24 / +15	+15 / 0	+32 / +23	6	10
10	18	+110 / 0	−95 / −205	+43 / 0	−50 / −120	+43 / 0	−32 / −75	+27 / 0	−16 / −34	+18 / 0	−6 / −17	+18 / 0	−11 / 0	+18 / 0	+12 / +1	+18 / 0	+23 / +12	+18 / 0	+29 / +18	+18 / 0	+39 / +28	10	18
18	30	+130 / 0	−110 / −240	+52 / 0	−65 / −149	+52 / 0	−40 / −92	+33 / 0	−20 / −41	+21 / 0	−7 / −20	+21 / 0	−13 / 0	+21 / 0	+15 / +2	+21 / 0	+28 / +15	+21 / 0	+35 / +22	+21 / 0	+48 / +35	18	30
30	40	+160 / 0	−120 / −280	+62 / 0	−80 / −180	+62 / 0	−50 / −112	+39 / 0	−25 / −50	+25 / 0	−9 / −25	+25 / 0	−16 / 0	+25 / 0	+18 / +2	+25 / 0	+33 / +17	+25 / 0	+42 / +26	+25 / 0	+59 / +43	30	40
40	50	+160 / 0	−130 / −290																	+25 / 0	+59 / +43	40	50
50	65	+190 / 0	−140 / −330	+74 / 0	−100 / −220	+74 / 0	−60 / −134	+46 / 0	−30 / −60	+30 / 0	−10 / −29	+30 / 0	−19 / 0	+30 / 0	+21 / +2	+30 / 0	+39 / +20	+30 / 0	+51 / +32	+30 / 0	+72 / +53	50	65
65	80	+190 / 0	−150 / −340																	+30 / 0	+78 / +59	65	80
80	100	+220 / 0	−170 / −390	+87 / 0	−120 / −260	+87 / 0	−72 / −159	+54 / 0	−36 / −71	+35 / 0	−12 / −34	+35 / 0	−22 / 0	+35 / 0	+25 / +3	+35 / 0	+45 / +23	+35 / 0	+59 / +37	+35 / 0	+93 / +71	80	100
100	120	+220 / 0	−180 / −400																	+35 / 0	+101 / +79	100	120
120	140	+250 / 0	−200 / −450	+100 / 0	−145 / −305	+100 / 0	−84 / −185	+63 / 0	−43 / −83	+40 / 0	−14 / −39	+40 / 0	−25 / 0	+40 / 0	+28 / +3	+40 / 0	+52 / +27	+40 / 0	+68 / +43	+40 / 0	+117 / +92	120	140
140	160	+250 / 0	−210 / −460																	+40 / 0	+125 / +100	140	160
160	180	+250 / 0	−230 / −480																	+40 / 0	+133 / +108	160	180
180	200	+290 / 0	−240 / −530	+115 / 0	−170 / −355	+115 / 0	−100 / −215	+72 / 0	−50 / −96	+46 / 0	−15 / −44	+46 / 0	−29 / 0	+46 / 0	+33 / +4	+46 / 0	+60 / +31	+46 / 0	+79 / +50	+46 / 0	+151 / +122	180	200
200	225	+290 / 0	−260 / −550																	+46 / 0	+159 / +130	200	225
225	250	+290 / 0	−280 / −570																	+46 / 0	+169 / +140	225	250
250	280	+320 / 0	−300 / −620	+130 / 0	−190 / −400	+130 / 0	−110 / −240	+81 / 0	−56 / −108	+52 / 0	−17 / −49	+52 / 0	−32 / 0	+52 / 0	+36 / +4	+52 / 0	+66 / +34	+52 / 0	+88 / +56	+52 / 0	+190 / +158	250	280
280	315	+320 / 0	−330 / −650																	+52 / 0	+202 / +170	280	315
315	355	+360 / 0	−360 / −720	+140 / 0	−210 / −440	+140 / 0	−125 / −265	+89 / 0	−62 / −119	+57 / 0	−18 / −54	+57 / 0	−36 / 0	+57 / 0	+40 / +4	+57 / 0	+73 / +37	+57 / 0	+98 / +62	+57 / 0	+226 / +190	315	355
355	400	+360 / 0	−400 / −760																	+57 / 0	+244 / +208	355	400
400	450	+400 / 0	−440 / −840	+155 / 0	−230 / −480	+155 / 0	−135 / −290	+97 / 0	−68 / −131	+63 / 0	−20 / −60	+63 / 0	−40 / 0	+63 / 0	+45 / +5	+63 / 0	+80 / +40	+63 / 0	+108 / +68	+63 / 0	+272 / +232	400	450
450	500	+400 / 0	−480 / −880																	+63 / 0	+292 / +252	450	500

All tolerance values are in 0.001 mm.

Fig. 6.24 British Standard data sheet BS 4500A: selected ISO fits – hole basis

British Standard BS 4500, *ISO limits and fits*, gives a selection of hole and shaft tolerances to cover a wide range of engineering applications.

For a selected range of fits which is adequate for most practical requirements, the BS 4500A and BS 4500B data sheets give the fits on a hole and shaft basis respectively.

For most general applications the hole-basis fits are usually recommended. Data sheet BS 4500A, fig. 6.24, shows a range of fits derived from selected hole tolerances (H11, H9, H8, H7) and shaft tolerances (c11, d10, e9, f7, g6, h6, k6, n6, p6, s6), where capital letters refer to holes, lower-case letters to shafts, and greater numbers to bigger tolerances.

Determining working limits

We will use the data sheet BS 4500A, fig. 6.24, to determine the working limits for the assembly shown in fig. 6.25(a), assuming this to be a designer's layout which has been passed to a detail draughtsman for preparation of the working drawings of the components.

In order to decide on a desirable fit, we must consider the function of the assembly. The shaft is going to rotate in the bush; hence a clearance fit is required, allowing sufficient space for lubricant, but not so much as to cause wobbling of the shaft. The most suitable fit will be H8/f7, as shown in fig. 6.25(b).

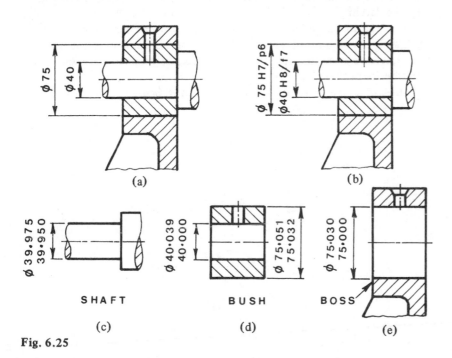

Fig. 6.25

We now locate 40 mm in the column of nominal sizes in the data sheet BS 4500A in fig. 6.24, remembering that column 'Over' means 'over but excluding' and 'To' means 'to and including'.

The required tolerances are shown below:

Over	To	H8	f7
30	40	+39	−25
		0	−50

As all tolerances are given in micrometres (0.000 001 m or 0.001 mm), the shaft basic size 40 f7 will have

maximum limit   40.000 − 0.025 = 39.975 mm
and   minimum limit   40.000 − 0.050 = 39.950 mm

as shown in fig. 6.25(c).

(When tolerancing, the same number of decimal places must be used for both limits.)

The bush hole basic size 40 H8 will have

maximum limit   40.000 + 0.039 = 40.039 mm
and   minimum limit   40.000 + 0   = 40.000 mm

as shown in fig. 6.25(d).

Assuming that the bush is going to be pressed into the bracket boss, then an interference fit will be appropriate. The h7/p6 fit seems to be most suitable, as shown in fig. 6.25(b).

Now locate 75 mm in the column of nominal sizes in the data sheet BS 4500A, fig. 6.24:

Over	To	H7	p6
65	80	+30	+51
		0	+32

The bush outside-diameter basic size 75 p6 will have

maximum limit   75.000 + 0.051 = 75.051 mm
and   minimum limit   75.000 + 0.032 = 75.032 mm

as shown in fig. 6.25(d).

The bracket boss basic size 75 H7 will have

maximum limit   75.000 + 0.030 = 75.030 mm
and   minimum limit   75.000 + 0   = 75.000 mm

as shown in fig. 6.25(e).

A toleranced drawing of a rectangular component is shown in fig. 6.26(a) and how it is interpreted is shown in fig. 6.26(b). The tolerance zones of 0.6 mm shown are the differences between the maximum and minimum limits.

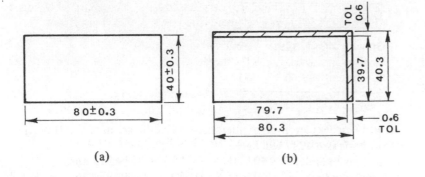

Fig. 6.26

Dimensioning tolerances

Tolerance limits between centres of holes can be indicated either by *chain dimensioning*, as in fig. 6.27, or by *progressive dimensioning* from a common datum as shown in fig. 6.28.

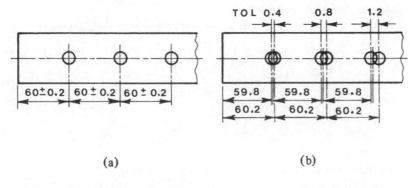

Fig. 6.27   Chain dimensioning

The use of chain dimensions results in an accumulation of tolerances between the holes and the edge of the plate, and this may endanger the functional requirements.

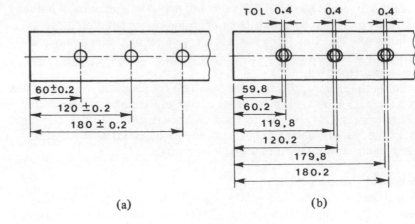

Fig. 6.28   Progressive dimensioning

Progressive dimensioning from a common datum on the component prevents this accumulation of tolerances, as each hole is toleranced directly from the datum.

When tolerancing an individual linear dimension, the method of specifying directly maximum and minimum limits of size is preferable. The larger limit should be given first, and the same number of decimal places should be indicated for both limits.

Four correct methods of tolerancing are shown below:

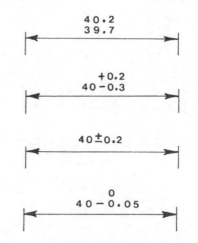

6.9 Machining symbol

An engineering component may be cast, forged, stamped, drawn, welded, etc., and after any of these processes it is usual for some surfaces to be machined. It is important that these surfaces are clearly indicated on the drawing.

The standard of finish or the degree of roughness of surface is very important in engineering and depends on the function and wear of the component. A very high degree of surface finish is usually required for precision work, for rotating and sliding parts, and for location purposes. Smooth surfaces are expensive to produce, and designers should keep the surface requirements to a minimum.

If any machined surface a section through is magnified many times, it will look like a range of mountains. The roughness of the surface is determined by measuring the distance in micrometres between the 'valleys' and the 'peaks'.

To indicate the surfaces to be machined, the symbol shown in fig. 6.29(a) should be used. The symbol should be applied normal to the line representing the surface to be machined, or it may be applied to a leader or extension line as in fig. 6.29(b).

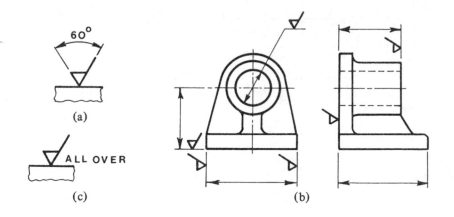

Fig. 6.29   Machining symbol

Where all the surfaces are to be machined, a general note may be used as shown in fig. 6.29(c).

6.10 Test questions

1. Explain very briefly the difference between size dimensions and location dimensions.
2. With the help of simple sketches, define the following types of dimension: (a) functional, (b) non-functional, (c) auxiliary (redundant).
3. Which of the following statements are true and which are false?

   a) Leaders end in dots, when crossing the outline.
   b) All radius dimension lines should pass through or be in line with the centre of the arc.
   c) Leaders are never used for dimensioning diameters.
   d) When a leader touches a circle, it should not be in line with the centre of the arc.
   e) Dimensions not drawn to scale should be put in brackets.
   f) Inspection dimensions always include tolerances.
   g) An assembly drawing never includes any dimensions.

4. Dimension the sectional view of the bracket shown in fig. 6.30 for inspection purposes only. The holes are to have H11 fits and the linear tolerances are to be 0.1 mm. Tracing paper may be used.

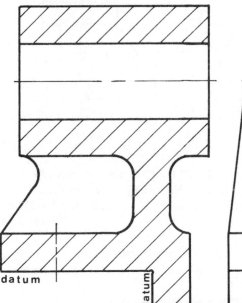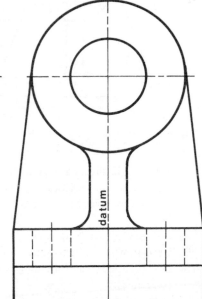

Fig. 6.30

5. Add a plan view and dimension the bracket shown in fig. 6.30 for casting, including all necessary instructions. Tracing paper may be used.
6. The bracket shown in fig. 6.31 is drawn to half full size (1:2).

   Redraw the bracket full size and dimension it for casting, including all necessary instructions. (Alternatively, tracing paper may be used for a half-full-size drawing.)

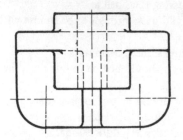

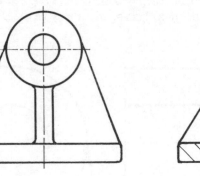

Fig. 6.31

7. Figure 6.32 shows a fully dimensioned template. Assume all numerical values to be in mm and to be correct. How many dimensions are incorrectly given with respect to BS 308–10? 11? 12? 13? 14? 15? or 16?
8. Define the term 'tolerance'.
9. Explain why tolerances are used for dimensioning purposes.
10. Discuss briefly the following considerations in establishing reasonable tolerances:

    a) function,

    b) interchangeability,

    c) cost.

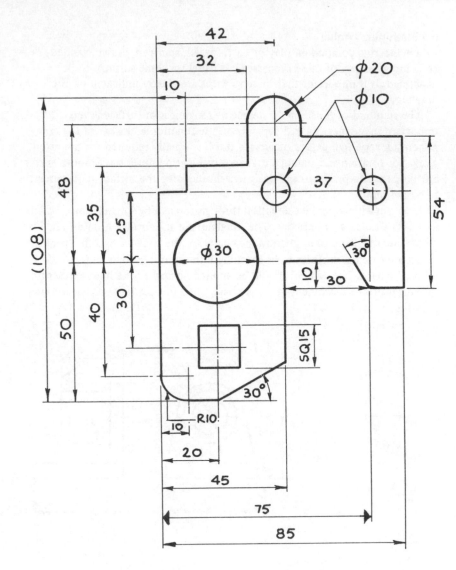

Fig. 6.32

11. Define the following terms as specified by BS 4500:

    a) basic size,

    b) deviation,

    c) maximum limit of size.

12. Describe the following types of engineering fits, and give two typical applications for each:

    a) clearance,
    b) transition,
    c) interference.

13. Explain, giving reasons, why the hole-basis system of fits is generally preferable to the shaft-basis system.

14. Calculate the maximum and minimum limits for the following shaft and hole nominal sizes:

    a)   50 H11/c11,
    b) 100 H7/n6,
    c) 150 H7/s6.

15. With the help of simple sketches, show how limits between the centres of holes can be indicated by

    a) chain dimensioning,
    b) progressive dimensioning.

Identify the main disadvantage of chain dimensioning.

16. Explain the meaning of the term 'datum' and give examples of two different features used as a datum.

17. What indication is given on a drawing that all surfaces of a component must be machined?

18. The limits between centres of four holes A, B, C, and D are indicated by chain dimensioning in mm:

<table>
<tr><td>limits between holes A and B are</td><td>20.02</td></tr>
<tr><td></td><td>19.98</td></tr>
<tr><td>limits between holes C and D are</td><td>30.02</td></tr>
<tr><td></td><td>29.98</td></tr>
<tr><td>limits between holes A and D are</td><td>60.00</td></tr>
<tr><td></td><td>59.88</td></tr>
</table>

Calculate the limits between holes B and C if all holes lie in sequence along the same centre line.

19. Fully dimension the component shown in fig. 6.33. Choose your own datum, include machining symbols and the cutting plane, and indicate the angle of projection.

20. Fully dimension the component shown in fig. 6.34. Tracing paper may be used.

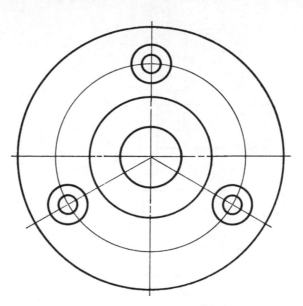
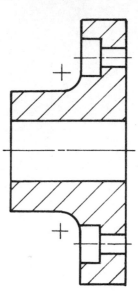

Fig. 6.33

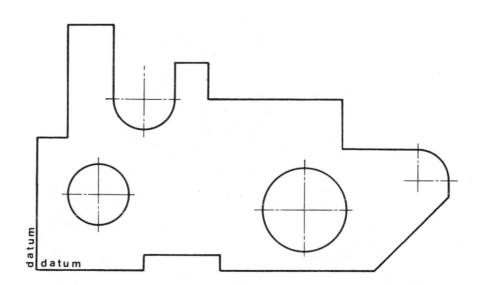

Fig. 6.34

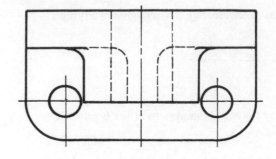

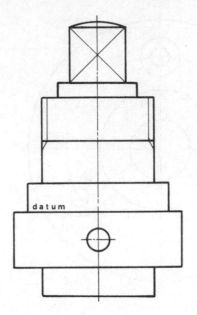

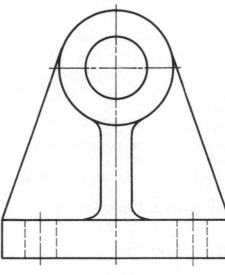

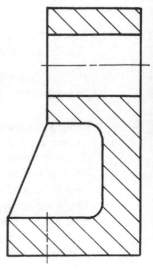

datum

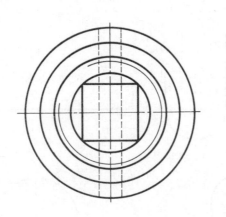

Fig. 6.36

Fig. 6.35

21. Fully dimension the component shown in fig. 6.35 and include the projection symbol. Tracing paper may be used.
22. The bracket shown in figure 6.36 is drawn to half full size (1:2).
    Redraw the bracket full size, and dimension it for machining processes. Include all required machining symbols, and tolerance the horizontal hole using a H9 fit for a 32 mm diameter nominal size. (Alternatively, tracing paper may be used for a half-full-size drawing.)

23. Figure 6.37 shows a partially dimensioned template. Assume all numerical values to be in mm and to be correct.
    How many dimensions are correctly given with respect to BS 308– 0? 1? 2? 3? 4? 5? or 6?
24. Fully dimension the component shown in fig. 6.38. Choose your own datum, include machining symbols, and indicate three functional dimensions by using a letter F.

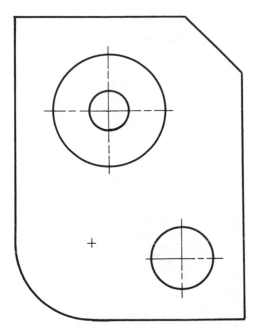

Fig. 6.37

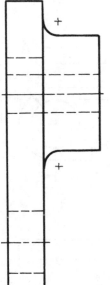

Fig. 6.38

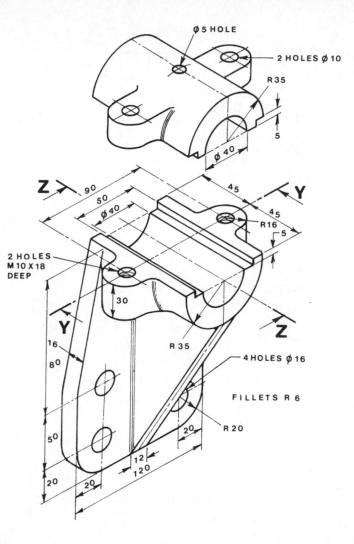

Fig. 6.39

25. Draw the following views of the assembled parts shown in fig. 6.39, adding studs, nuts, washers, and a 120 mm length of 40 mm diameter shaft:
   a) a sectional front view on YY,
   b) a sectional end view on ZZ,
   c) a plan view showing hidden detail.
   Include the functional dimensions, insert a title block, and add a parts list.

71

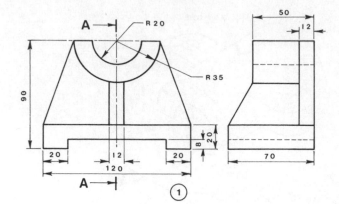

①

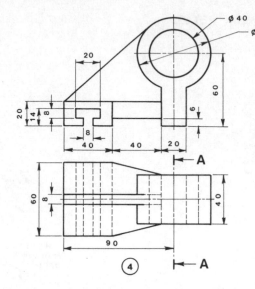

④

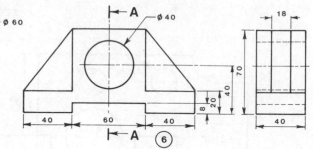

②

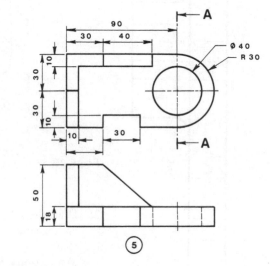

⑤

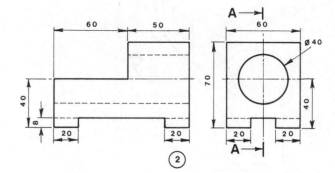

③

⑥

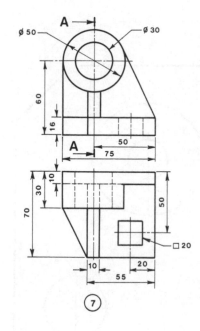

⑦

Fig. 6.40

72

26. Draw a sectional front view on AA, a plan view, and an end view of each component 1 to 7 shown in fig. 6.40. Components 1, 2, 3, and 5 are to be drawn in first-angle projection and components 4, 6, and 7 in third-angle projection and fully dimensioned.

27. Draw in oblique projection components 3 and 7 shown in fig. 6.40.

28. Draw in isometric projection components 5 and 6 shown in fig. 6.40.

29. Draw full size in third-angle projection the following views of the compressor crosshead shown in fig. 6.41:
    a) a sectional front view on AA,
    b) a sectional plan view on BB,
    c) an end view showing hidden detail.
    Dimension the crosshead for inspection purposes only.

30. Draw full size in first-angle projection the following views of the bracket shown in fig. 6.42:
    a) a sectional front view on AA,
    b) an end view in the direction of arrow B. Include all hidden detail in this view.

    You may need to construct part of a plan view to enable the front view to be drawn correctly.

    Dimension the bracket for machining purposes only.

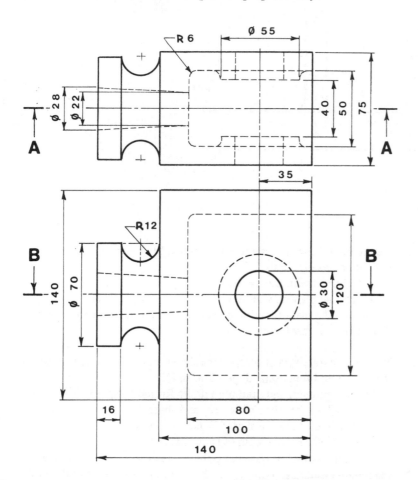

Fig. 6.41

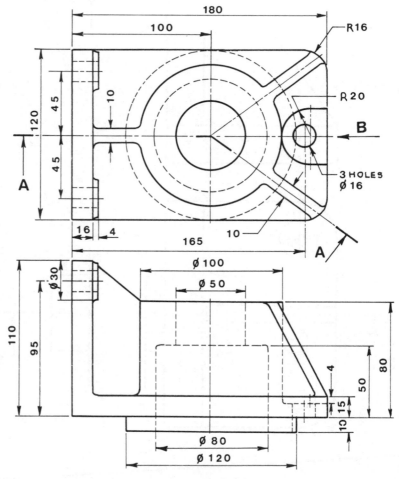

FILLETS 4 RAD

Fig. 6.42

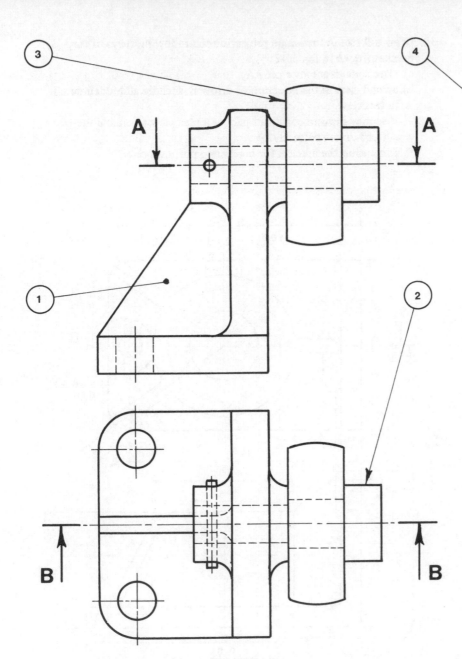

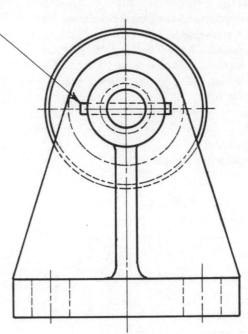

31. Figure 6.43 shows an assembly of a bracket (1), shaft (2), and pulley (3) with a pin (4). Draw
   a) a sectional front view on BB,
   b) a sectional plan view on AA.
      Include the main functional dimensions and the projection symbol. Tracing paper may be used.

32. Redraw separately each component shown in fig. 6.43 and include the inspection and functional dimensions only.
      Show maximum and minimum limits for the following fits:
      14 H8/f7   between pulley (3) and shaft (2),
      10 H7/k6   between shaft (2) and bracket (1),
       3 H7/p6   between pin (4) and shaft (2).
      The longitudinal fit between pulley (3) and pin (4) is to have a maximum clearance of 0.8 mm and a minimum clearance of 0.4 mm.
      Tracing paper may be used.

Fig. 6.43

74

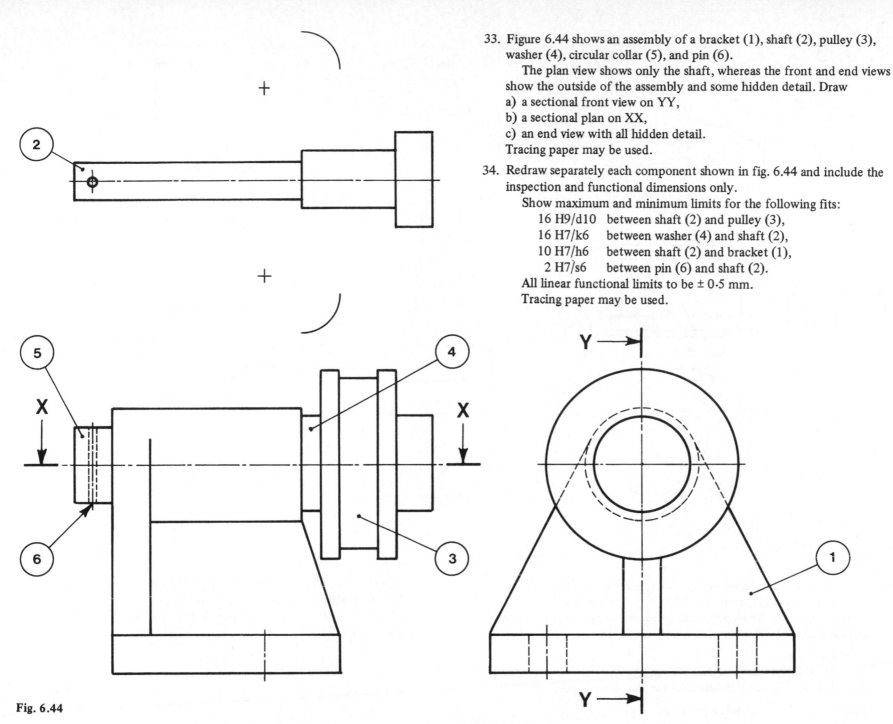

33. Figure 6.44 shows an assembly of a bracket (1), shaft (2), pulley (3), washer (4), circular collar (5), and pin (6).

    The plan view shows only the shaft, whereas the front and end views show the outside of the assembly and some hidden detail. Draw
    a) a sectional front view on YY,
    b) a sectional plan on XX,
    c) an end view with all hidden detail.
    Tracing paper may be used.

34. Redraw separately each component shown in fig. 6.44 and include the inspection and functional dimensions only.

    Show maximum and minimum limits for the following fits:
      16 H9/d10   between shaft (2) and pulley (3),
      16 H7/k6   between washer (4) and shaft (2),
      10 H7/h6   between shaft (2) and bracket (1),
       2 H7/s6   between pin (6) and shaft (2).
    All linear functional limits to be ± 0·5 mm.
    Tracing paper may be used.

Fig. 6.44

75

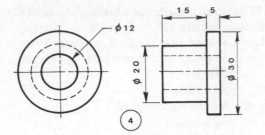

4

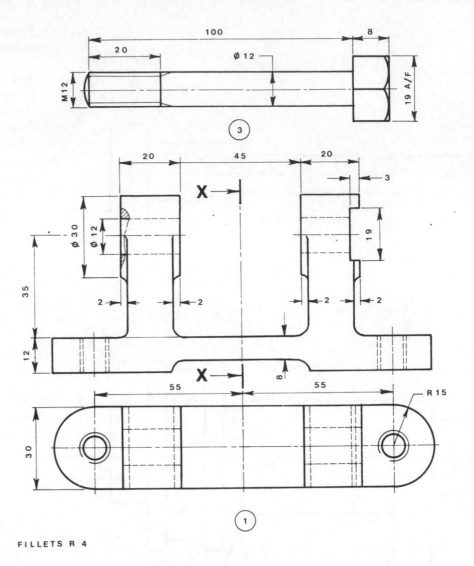

3

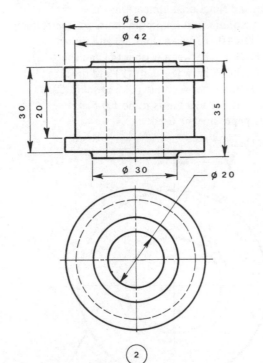

2

1

FILLETS R 4

Fig. 6.45

35. The belt-pulley unit shown in fig. 6.45 consists of a belt pulley (2), a mounting bracket (1), a fitted bolt (3), and two bushes (4).

    Draw full size with all parts assembled, including a suitable nut with locking device (see chapter 7),

  a) a front view,
  b) a sectional end view on XX,
  c) a plan view. Show all hidden detail in this view.

Insert a title block and add a parts list.

Dimension for inspection purposes only, with the following fits:
  H7/p6 between shaft (3) and two bushes (4),
  H9/e9 between two bushes (4) and belt pulley (2).
Linear tolerances to be ± 0.2 mm.

7 Fasteners

7.1 Screw threads

A knowledge of screw threads and fasteners is important when preparing working drawings.

A screw thread is a helical groove which is cut around a cylindrical external surface or in a cylindrical hole. The cylinder is then called a screw and the part with the hole a nut.

Screw threads may be right-hand or left-hand, depending on the direction of the helix. This can be represented by heavy strings wound round a rod as shown in fig. 7.1. A right-hand thread advances into a threaded hole when turned clockwise; a left-hand thread advances when turned anticlockwise.

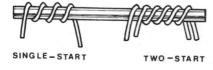
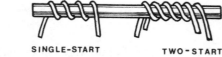

(a) Left-hand   (b) Right-hand

Fig. 7.1 Left-hand and right-hand threads

When a quick advance is desired, two or more threads are cut side by side and the thread is said to be 'two-start' etc.

There are two main types of thread form: *vee thread* and *square thread*, fig. 7.2.

(a) Vee thread   (b) Square thread

Fig. 7.2 Main types of thread form

The basic ISO (International Organisation for Standardisation) screw-thread form is shown in fig. 7.3.

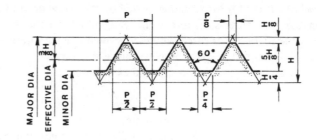

Fig. 7.3 ISO metric screw-thread form

Figure 7.4(a) shows internal and external threads. Figure 7.4(b) shows the conventional representation of screw threads, which is the same for all types of thread.

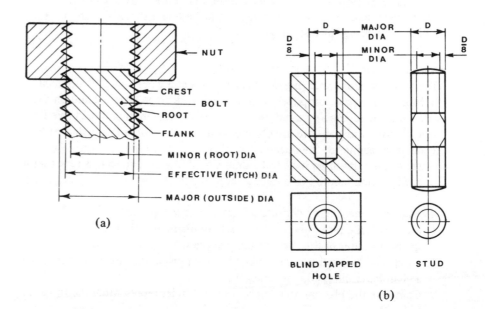

Fig. 7.4 (a) Internal and external threads
          (b) Conventional representation of screw threads

The *pitch* (*p*) of a thread is the distance, measured parallel to the axis, from any point on a thread to the corresponding point on the adjacent thread (see fig. 7.3). For a single-start thread it will equal the distance moved by the screw or nut during one revolution.

The *lead* of a thread is the distance moved by the screw or nut in one revolution; e.g. for a three-start thread the lead will be three times the pitch.

The *effective diameter* or *pitch diameter* is the diameter where the width of the tooth is equal to the space between successive teeth.

ISO metric screw threads are designated as follows:

M   8 x 1.25
                └──── Pitch in millimetres
           └──────── Nominal size in millimetres (major diameter)
  └─────────────── Symbol for ISO metric thread

7.2 Temporary fastenings

There are different types of fastening to join one part to another. The design and the function of the secured parts must be taken into consideration before finally deciding on the fastening method to be adopted.

There are two classes of fastenings: temporary and permanent.

Temporary fastenings can be used more than once. Any assembled components held together can be dismantled and re-assembled many times without damaging the fastenings.

Nuts and bolts

The bolt has an external thread which extends along only part of the shank. Bolts generally pass completely through the work to be fastened and on the other side are secured by a nut, which has an internal mating thread (see fig. 7.9(a)). Nuts and bolts are usually hexagonal-headed and are adjusted with a standard spanner of the open-ended, ring, or socket type.

Nuts and bolts often have to be drawn by draughtsmen, so it is very useful to learn a quick method to obtain an approximate shape.

Approximate method of drawing a hexagonal nut (fig. 7.5)

1. Start with the plan view. Draw a circle of diameter 2*D*, where *D* is the major diameter of the thread (nominal size).
2. Using a 60° set square, construct a hexagon inside the circle and then draw a chamfer circle inside the hexagon.
3. Complete the plan by drawing concentric circles representing the threaded hole.
4. Project the front and end views, making the height of the nut equal to *D*.

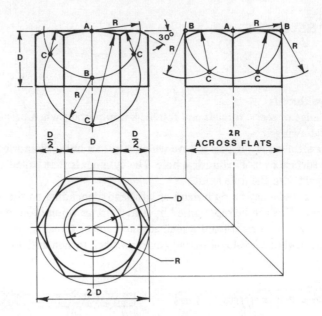

Fig. 7.5 Method of drawing a hexagonal nut

5. With centres at points A, draw arcs of radius *R*, where *R* is the radius of the chamfer circle.
6. With centres at points B, draw arcs of radius *R* to obtain the intersection points C.
7. With centres at points C, draw chamfer curves tangential to the top surface of the nut.
8. Complete the views.

Figure 7.6 shows the completed views of a nut and bolt. Note the nut and bolt head heights and the full-thread and bolt lengths.

An alternative method of drawing chamfer curves is to use a 60° set square to draw construction lines from the points marked A in fig. 7.7. The intersection points marked B are the required centres for the chamfer curves tangential to the top surface of the nut, as shown.

To obtain an end view of the nut, draw an overall rectangular shape projected from the plan via a deflector as shown. Then draw the required tangential arcs from the mid-points B with radii equal to the major diameter *D*.

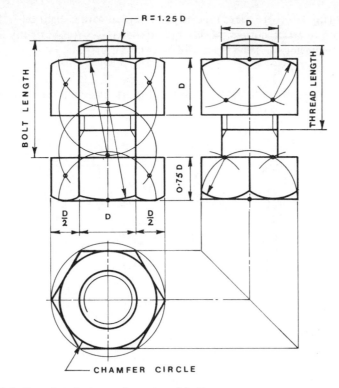

Fig. 7.6 Completed views of a nut and bolt

Fig. 7.7 Alternative method of drawing a nut

Screws

A screw has an external thread extending almost the whole length of the screw shank to the head.

Screws, like bolts, are used for fastening two or more parts together. One of the parts has a tapped hole and the other part has a clearance hole (see fig. 7.9(b)). The screw is used by passing it through the clearance hole in one part to screw into the threaded hole in the other, so fastening both parts securely together.

Screws are not secured by a nut.

Apart from hexagonal heads, the following types of screw head are the most regularly used:

a) round head,
b) cheese head,
c) countersunk head,
d) socket head.

For socket-head screws a hexagonal-bar spanner or Allen key is used; for the remainder a screwdriver is used.

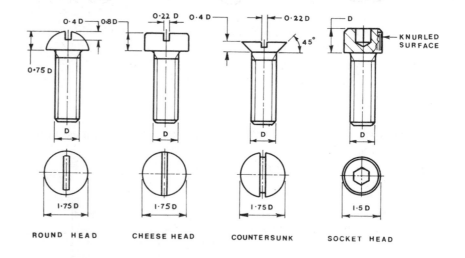

Fig. 7.8 Common types of screw

Figure 7.8 shows the common types of screw with all proportions indicated in terms of the nominal size, i.e. the major diameter of the thread (*D*).

Studs

Studs are threaded on both ends, with an unthreaded shank in the middle, and are used for parts that must be removed frequently, like cylinder heads, covers, lids, etc.

Studs are screwed tightly into tapped holes in the permanent part, while the removable part has clearance holes in the corresponding positions. Nuts are used on the projecting ends of the studs to secure the two parts together, as shown in fig. 7.9(c).

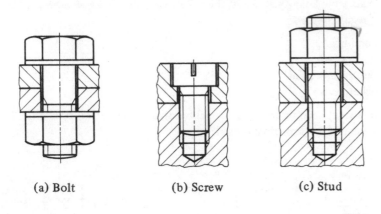

(a) Bolt   (b) Screw   (c) Stud

Fig. 7.9

Pins

Dowel pin (fig. 7.10(a))   A dowel pin is a headless cylindrical pin used for precise-location purposes.

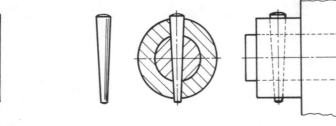

(a) Dowel pin   (b) Taper pin

Fig. 7.10

Taper pin (fig. 7.10(b))   This type of pin is conical with a slight taper. It is usually used to attach cotters, wheels, etc. to shafts. It is forced tightly into a reamed hole having the same taper, which is standardised.

Split pin   The split pin is usually inserted through holes and slots and its ends are opened up as in fig. 7.11.

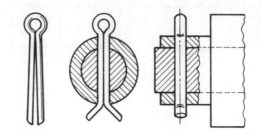

Fig. 7.11   Split pin

Cotter pin   The cotter pin is a round rod threaded at one end, or it may be plain with a tapered flat machined along its length. This type is used to secure levers, cranks, etc. to spindles, fig. 7.12. A typical application is attaching a bicycle-pedal crank to the chainwheel spindle.

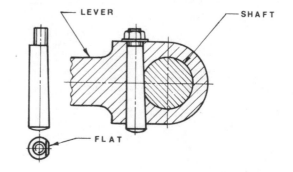

Fig. 7.12   Cotter pin

Circlips

A component can be located axially against a shoulder of a shaft by a circlip or snap ring as in fig. 7.13. To provide axial location between a shaft and its bearing, a locating groove is machined and the circlip is sprung into position. Circlips may be internal or external.

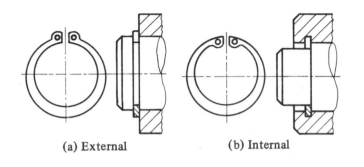

(a) External   (b) Internal

Fig. 7.13   Circlips

7.3 Permanent fastenings

Permanent fastening methods involve the destruction of the fastening if the components ever need to be separated once they have been joined.

Rivets

Rivets are permanent fasteners and are used for joining metal sheets and plates in structural shipbuilding, boiler and aircraft work, etc.

Rivets are cylindrical rods with one head formed when manufactured. A head is formed on the other end when the parts to be joined have been assembled and the rivet has been placed through the holes of the mating parts. The holes should have a good clearance, so that the rivet can swell during the forging of the head.

Rivets are made of a ductile material such as low-carbon steel, brass, copper, or duralumin.

Simple riveting operation   Suitable holes are drilled or punched in the parts which are to be joined together. The rivets, which may be cold or heated, are placed in the holes and are forged to the required shape using either a hand hammer or a pneumatically or hydraulically operated gun (fig. 7.14).

The different types of rivet are defined according to the shape of the rivet head.

Snap- or round-head rivets are the most commonly used and are easy to shape. These rivets are used mainly for machine riveting and structural work and are riveted over to the same shape at the opposite end, fig. 7.14.

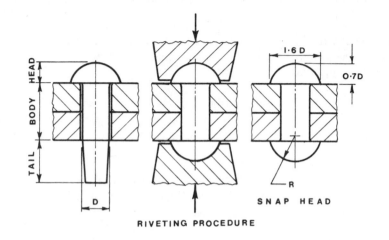

Fig. 7.14   Snap-head rivet

Pan-head rivets are considered to be very strong and are usually riveted over to a snap head at the opposite end, fig. 7.15(a).

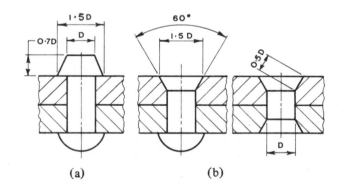

Fig. 7.15   (a) Pan-head and (b) countersunk-head rivets

Countersunk-head rivets, fig. 7.15(b), are mainly used in shipbuilding or where flush surfaces are required. Sometimes they are riveted over to the same shape at the opposite end, hence leaving the rivets flush with the work surface on both sides.

Welding

Welding is a process of uniting two pieces of metal by fusing them together to form a permanent joint. This may be done with or without additional (filler) metal and with or without the application of pressure.

Fusion welding   In this process where the areas to be joined are heated until they become plastic (i.e. able to flow) and are then welded together with or without the addition of filler metal.

Gas welding is a fusion process in which the heat is provided by burning a gas mixed with oxygen to create a hot flame which is applied to the joint by means of a torch. The most commonly used gases are acetylene, propane, and hydrogen.

Electric-arc welding is a fusion process in which a local area of intense heat is created by passing an electric current through a filler rod, which acts as an electrode, held at a short distance from the joint so that the electric circuit is completed by arcing.

Pressure welding   In this process the areas to be joined are heated until they become plastic and are then welded together by applying pressure or sometimes hammering.

In *forge welding*, the pieces to be joined are heated and then hammered together.

In *resistance welding*, the pieces to be joined are butted together under pressure and a heavy current is passed through them, producing sufficient heat for welding under continual pressure.

Spot welding is a process of pressure resistance welding in which thin parts are overlapped and welds are made at successive single spots by electrodes contacting both sides of the metal sheets under pressure.

Welding symbols

On engineering drawings, welding requirements are made clear and unambiguous by using welding symbols specified in British Standard BS 499. These symbols give instructions as to the type of welds, their position, and sometimes their size.

The type of weld to be made is indicated by the type of weld symbol. Some of the welds commonly used and their symbols are shown in fig. 7.16.

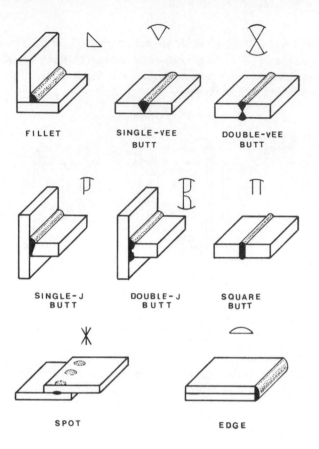

Fig. 7.16   Welds and symbols

The method used to indicate a weld on a drawing is shown in fig. 7.17:

a) if the weld symbol is below the horizontal reference line, the weld is to be made on the arrow side of the joint;

b) if the weld symbol is above the reference line, the weld is to be made on the side of the joint away from the arrow, known as the 'other' side;

c) if the weld symbol is shown above and below the reference line, welds are to be made on both sides of the joint.

82

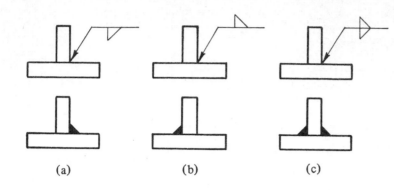

Fig. 7.17 Weld indication on drawings

A weld all round is indicated by a circle drawn at the elbow of the arrow, fig. 7.18(a), and a weld to be made on site is indicated by a filled-in circle as in fig. 7.18(b).

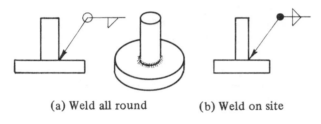

(a) Weld all round   (b) Weld on site

Fig. 7.18

If a weld requires special preparation, the arrowhead should always point to the edge of the plate.

7.4 Locking devices

In machinery which is subject to constant working vibration, nuts and bolts tend to work loose — often with disastrous results.

Many methods of locking nuts and bolts have been used, and these fall into two basic groups:

a) friction locking, which relies on increased friction between the nut and the bolt;
b) positive locking, which relies on the use of a pin through the nut, various locking plates, washers, or wire locks.

Friction locking devices

Friction locking is usually required on machines which are subject to light vibration. (Friction is the resistance to motion of one surface relative to another surface in contact.)

Lock nut Two nuts screwed firmly one on top of the other strain against each other and wedge the nut threads on to the opposite flanks of the bolt threads. This has a locking effect, fig. 7.19(a).

The thinner lock nut is sometimes fitted beneath the main nut, which carries all the tensile load, and sometimes on top, thus eliminating the need for a thin spanner to adjust it.

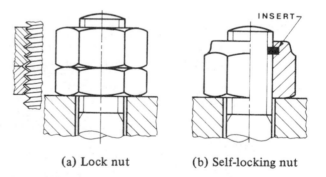

(a) Lock nut   (b) Self-locking nut

Fig. 7.19

Self-locking nuts These contain a fibre or plastics ring insert, fig. 7.19(b), which is compressed against the bolt thread when the nut is tightened, thus producing frictional gripping action.

Plain washer A plain washer is not a locking device but provides a flat base for a nut and stops the nut digging in. It is useful on rough surfaces and slots.

Spring washer   When a nut is tightened on to a spring washer, fig. 7.20(a), the washer's spring action pushes the nut threads against the bolt threads, so increasing the frictional forces.

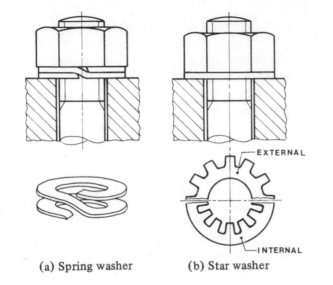

(a) Spring washer   (b) Star washer

Fig. 7.20

Star washer   These serrated washers have sharp corners, fig. 7.20(b), which dig into the nut, producing a wedging action and thus forming a semipositive lock.

Positive locking devices
Positive locking devices do not rely on friction and are usually required on machines which are subject to heavy vibrations.

Castle nut   A castle nut is a hexagonal nut with a cylindrical upper part in which slots are cut in line with the centre of each hexagonal face. A hole is drilled through the bolt to correspond with a pair of slots, then a split pin is inserted and the ends are bent over, fig. 7.21(a). Positive locking is achieved without reducing the number of full threads engaged with the bolt.

Tab washer   The tab washer is inserted between the component and the nut. After the nut has been tightened, one tab is bent closely against a flat of the nut, to prevent rotation, and the other tab is bent down into a drilled hole or over the edge of the component, fig. 7.21(b).

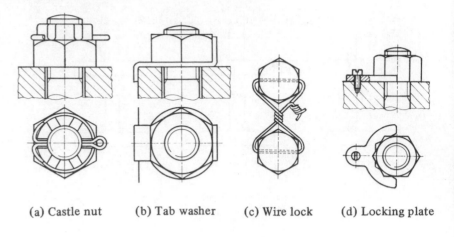

(a) Castle nut   (b) Tab washer   (c) Wire lock   (d) Locking plate

Fig. 7.21

Wire locking   In this method, soft wire is passed through holes drilled in adjacent bolt heads after the bolts have been tightened. The ends of the wire are twisted together with pliers. The wire must be so arranged that if a nut tends to slacken it should cause the wire to tighten, fig. 7.21(c).

Locking plate   A locking plate is screwed down so that its jaws engage with the nut and prevent it turning, fig. 7.21(d).

7.5 Test questions
1. Explain very briefly why washers are used in engineering.
2. Give the correct name for each of the washers shown in fig. 7.22. Which of them is used as a positive locking washer?

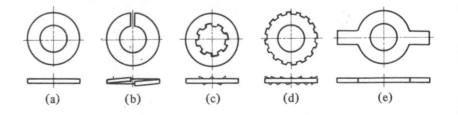

(a)   (b)   (c)   (d)   (e)

Fig. 7.22

3. Name the components shown in fig. 7.23 and sketch a simple assembly showing the practical application of these two components.

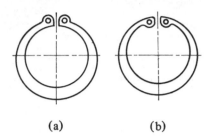

(a)   (b)

Fig. 7.23

4. Name all the indicated parts and features shown in the assembly in fig. 7.24.

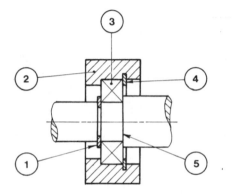

Fig. 7.24

5. a) What is the main difference between a screw and a bolt?
   b) What is the main difference between a bolt and a stud?
6. Identify the screw heads shown in fig. 7.25 by their correct names and suggest the tools used in fastening each of them.

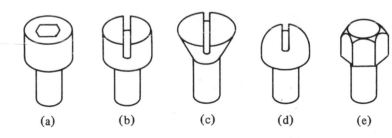

(a)   (b)   (c)   (d)   (e)

Fig. 7.25

7. Draw or sketch the fastenings indicated at (a), and (b), (c), and (d) in fig. 7.26 for joining two parts temporarily together. Name each fastener.

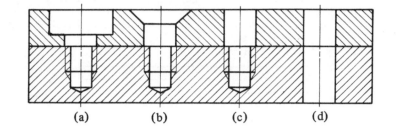

(a)   (b)   (c)   (d)

Fig. 7.26

8. Name the pins shown in fig. 7.27.

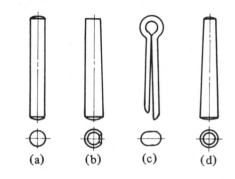

(a)   (b)   (c)   (d)

Fig. 7.27

9. The four plates shown in fig. 7.28 are not to be machined before welding. Sketch the sections of the most suitable types of weld to join these four plates together, and name each type chosen.

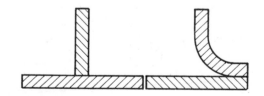

Fig. 7.28

85

10. Name the types of weld shown in fig. 7.29 and sketch the appropriate BS welding symbols inside the boxes provided.

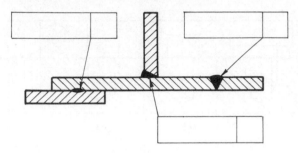

Fig. 7.29

11. By positioning the appropriate symbols on the reference lines, show how the welds in fig. 7.30 are indicated on a drawing.

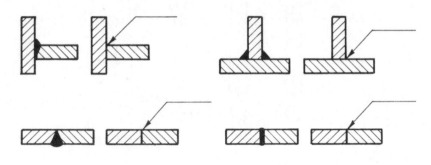

Fig. 7.30

12. State the correct names of the locking devices shown in fig. 7.31. Which of them is a positive locking device?

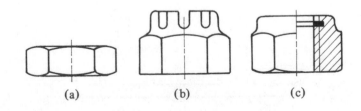

(a) (b) (c)

Fig. 7.31

13. State the correct names of the rivets shown in fig. 7.32. Which rivets are used for (a) structural work? (b) shipbuilding?

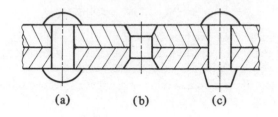

(a) (b) (c)

Fig. 7.32

14. State the meaning of the welding symbols shown in fig. 7.33 (a), (b), and (c).

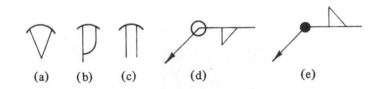

(a) (b) (c) (d) (e)

Fig. 7.33

15. Describe very briefly the meaning of the welding instructions indicated in fig. 7.33(d) and (e).

16. Explain the meaning of the terms 'temporary fastenings' and 'permanent fastenings' and name three examples of each.

17. State the meaning of the terms 'friction locking devices' and 'positive locking devices' and sketch three examples of each type.

18. Redraw, twice full size, the nut and bolt shown in fig. 7.34 and add a plan.

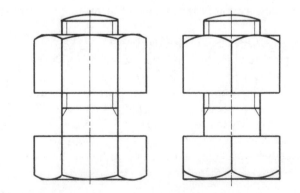

Fig. 7.34

Appendix: drawing ellipses

To draw an ellipse using two concentric circles (fig. A1)

1. Draw two concentric circles equal in diameter to the major axis XX and the minor axis YY of the required ellipse.

   Divide the circles into a number of parts with radial lines crossing the inner and outer circles.

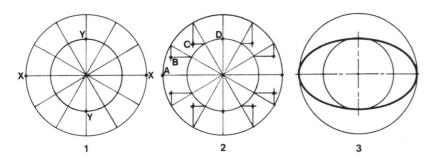

Fig. A1

2. Where the radial lines cut the inner and outer circles, draw horizontal and vertical lines respectively. The points of intersection A, B, C, and D are points on the ellipse.
3. Draw a uniform bold curve through the intersection points to form the required ellipse.

To draw an approximate ellipse using instruments (fig. A2)

1. Draw a construction rectangle with XX equal to the major axis and YY equal to the minor axis of the required ellipse.

   Bisect the angle A and the angle B, the bisectors intersecting at C.
2. With the centre at C, draw a circle tangential to XX and YY.

   From A, draw a tangent to the circle with centre C, intersecting YY at D and XX at E.

   By means of dividers or compasses and with the centre at B, using symmetry transfer E to E′ and D to D′.

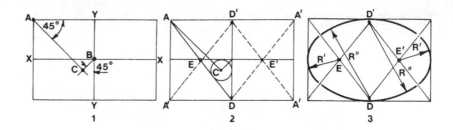

Fig. A2

3. With centres at E and E′, draw tangential arcs of radius R′, and with centres at D and D′ draw the remaining arcs of radius R″.

   (The points D and D′ may be outside the construction rectangle, depending on the ratio of axes.)

To draw an ellipse using the trammel method (fig. A3)

1. Draw the major axis AA and the minor axis BB. Mark off on the trammel (a strip of paper) half the length of the minor axis (EA).

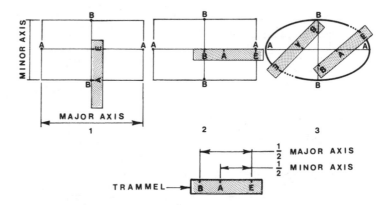

Fig. A3

2. Mark off on the trammel half the length of the major axis (EB), measured from the same end E.
3. Position the trammel on the drawing, making sure that point A always lies on the major axis AA and point B always lies on the minor axis BB. Mark a point E for each position of the trammel and join the points obtained with a smooth curve to form the required ellipse.

Selected solutions

Solutions to the descriptive questions can readily be found from the text; solutions to a selection of other questions are given here, though in some cases alternative correct solutions are possible.

All diagrams are drawn to a reduced scale, and, for simplicity, the fillets on certain drawings have been omitted.

Page 13, Q 20   (a) 9, (b) 8, (c) 2, (d) 13, (e) 3, (f) 10, (g) 4, (h) 14, (i) 11, (j) 5, (k) 6, (l) 1, (m) 7, (n) 12

Page 13, Q 21

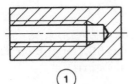
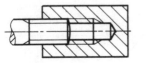
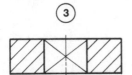

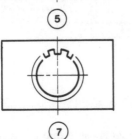
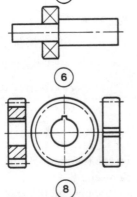

Page 21, Q 1

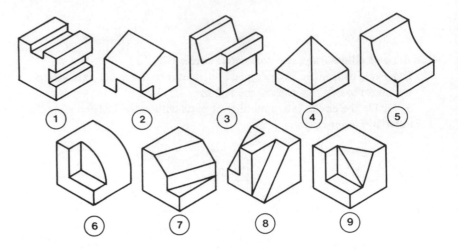

Page 22, Q 2

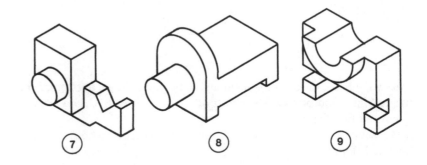

Page 25, Q 1

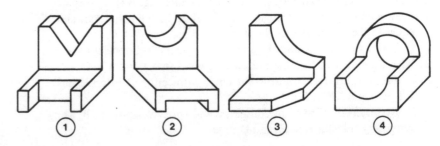

Page 28, Q 2

Page, 29, Q3

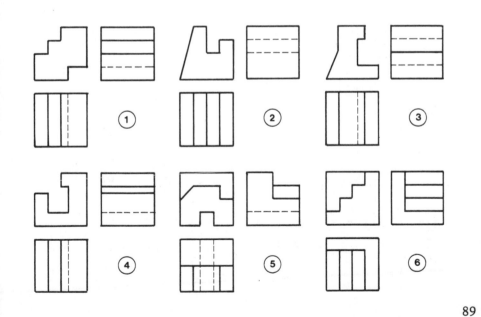

Page 32, Q 1

	A	B	C	D	E	F
FV	1	4	6	13	17	15
EV	18	7	16	2	3	12
PV	9	14	11	8	10	5

Page 33, Q2

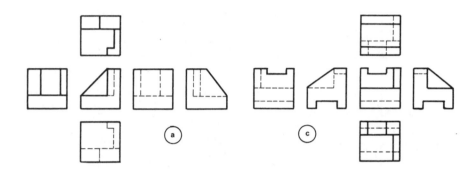

Page 33, Q 3

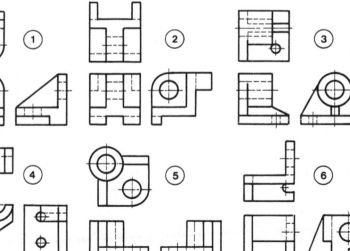

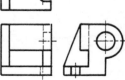

Page 34, Q 4

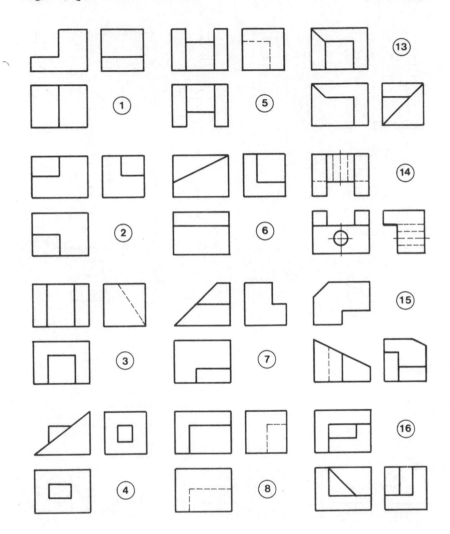

Page 35, Q5

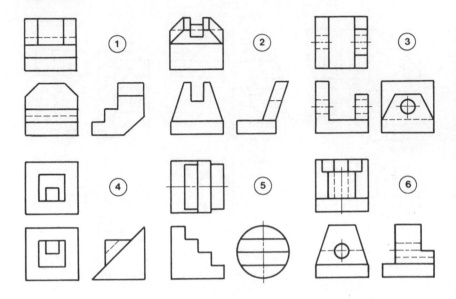

Page 41, Q 1

Page 36, Q 6   (a) B, (b) D, (c) E

Page 36, Q 7   (a) A, (b) D

Page 37, Q 9   (1) A, (2) B, (3) C, (4) D, (5) C, (6) A, (7) A, (8) C, (9) D

Page 42, Q 2

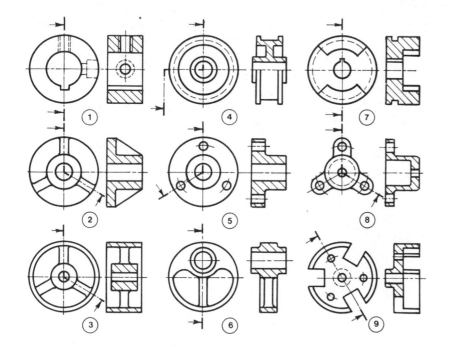

Page 44, Q 2

Page 43, Q 1

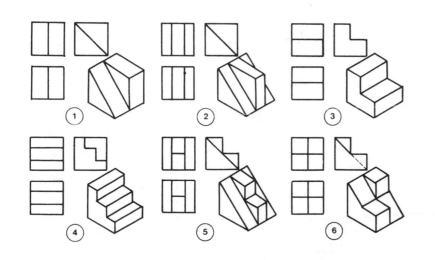

Page 47, Q1

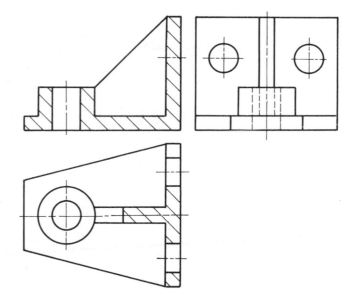

Page 53, Q 1

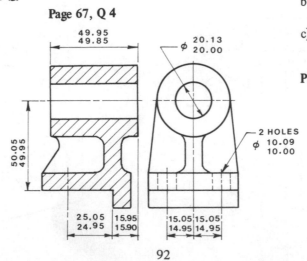

Page 54, Q 2

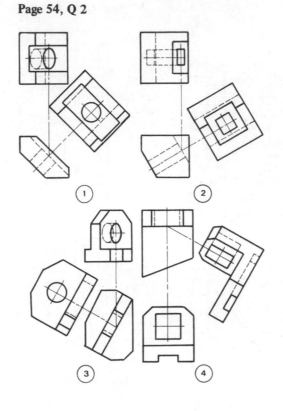

Page 68, Q 5

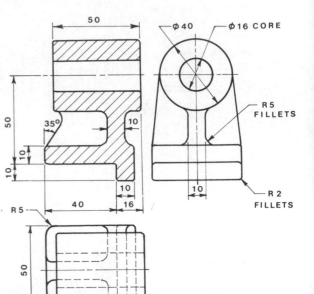

Page 55, Q 4   B

Page 55, Q 5   B

Page 67, Q 3   (a) T, (b) T, (c) F, (d) F, (e) F, (f) T, (g) F

Page 69, Q 14

a) hole	50.160	shaft	49.870
	50.000		49.710
b) hole	100.035	shaft	100.045
	100.000		100.023
c) hole	150.040	shaft	150.125
	150.000		150.100

Page 69, Q 18   9.96, 9.92

Page 67, Q 4

Page 69, Q 19

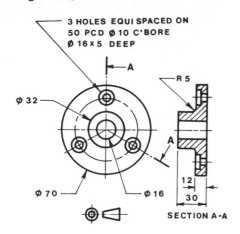

3 HOLES EQUI SPACED ON
50 PCD ∅ 10 C'BORE
∅ 16 × 5 DEEP

R 5

∅ 32

∅ 70

∅ 16

12

30

SECTION A-A

Page 69, Q 20

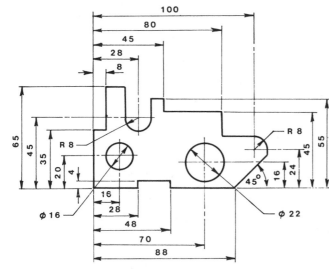

100

80

45

28

8

65

45

35

20

4

R 8

R 8

55

45

16

24

45°

16

28

48

70

88

∅ 16

∅ 22

Page 70, Q 21

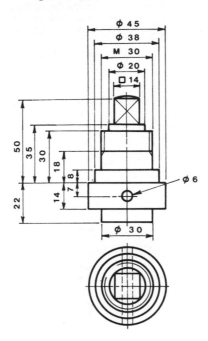

∅ 45

∅ 38

M 30

∅ 20

□ 14

50

35

30

18

8

22

14

7

∅ 6

∅ 30

Page 70, Q 23   1

Page 73, Q 30

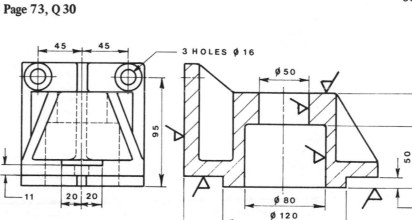

45

45

3 HOLES ∅ 16

∅ 50

95

∅ 80

∅ 120

100

80

50

10

11

20

20

Page 75, Q 33

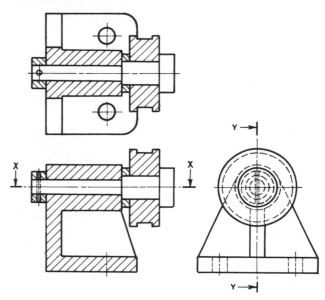

Y

X

X

Y

Page 86, Q 10

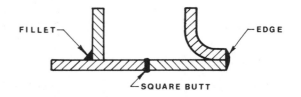

FILLET

EDGE

SQUARE BUTT

Page 85, Q 4

(1) external circlip,
(2) bearing housing,
(3) bearing,
(4) internal circlip,
(5) shaft shoulder.

Index

abbreviation, 7
angular dimensions, 58
assembly, 1, 2, 40, 62, 74, 75
auxiliary dimension, 56
auxiliary plane, 26, 49, 52
auxiliary view, 49, 52

basic size, 62, 63
bearing, 9, 13
bolt, 39, 40, 77–80
bore, 6
boss, 6
bush, 6

castle nut, 84
centre line, 4, 5, 7
chain dimension, 66
chamfer, 6, 7, 58
checker, 12
cheese head, 7, 79
chief draughtsman, 11
circlip, 81
clearance fit, 63, 65
collar, 6
communication, 1
component drawing, 1, 2
conventions, 8, 9
cotter pin, 80
counterbore, 6, 7, 59
countersink, 7, 59, 79, 81
cutting plane, 4, 38, 39, 40

datum, 56, 57
designer, 11
design layout, 2, 10
deviation, 62
die casting, 60
dimensioning, 56–66
dowel, 6, 80
draughtsman, 12
drawing, 1
drawing equipment, 2
drawing office, 10
drawing sequence, 46
drop forging, 60
dye-line process, 10

effective diameter, 78
ellipses, 87
engineering drawing, 1, 2

fabrication, 60, 61
fastening, 77–84
fillet, 6, 46
first-angle projection, 26–7
fitting dimension, 1, 2, 62
flange, 6, 13
form and proportion, 14
freehand sketching, 14
friction locking, 83
functional dimension, 56, 61, 62

general-arrangement drawing, 2

hatching, 4, 38
hexagon, 7, 78
hidden detail, 4, 5, 8
hole-basis system, 63
horizon, 15, 16
horizontal plane, 26, 31, 48, 50
hub, 6

inspection, 61
interference fit, 64
interrupted view, 8
isometric projection, 18, 20
ISO metric thread, 77

key, 6, 40
keyway, 6
knurling, 9

lead, 78
leader, 4, 57
lettering, 5
limits and fits, 62
lines, 4, 5
lines in space, 48–50
location dimension, 56
lock nut, 83
locking devices, 83

machining, 61
machining symbol, 67
major diameter, 77, 78
maximum limit, 62
microfilm process, 10
minimum limit, 62
minor diameter, 77

non-functional dimension, 56
nut, 19, 40, 78, 83

oblique projection, 23, 24
organisation, 10
orthographic projection, 26, 31
outline, 4

pan head, 81
partial view, 45
parts list, 2, 62
permanent fastening, 81
perspective projection, 15
pictorial projection, 15
pin, 80
pitch, 7, 9, 78
pitch-circle diameter, 9, 58
plain washer, 83
points in space, 48
positive locking devices, 84
primary production, 60
principal planes, 26, 31
principles of dimensioning, 56
print room, 10
progressive dimensioning, 66
projection lines, 57
projection symbol, 27, 31
projectors, 15

reading a drawing, 43
recess, 6
redundant dimension, 56, 59
rib, 6, 40
rim, 6
rivet, 81
round head, 7, 79, 81

sand casting, 60
scale, 3, 45
screw, 79, 80
screw thread, 7, 77
secondary production, 60
sectional views, 4, 38
section leader, 11
self-locking nut, 83

shaft, 6, 8, 39, 40
shaft-basis system, 63
single-part drawing, 1, 2
single-start thread, 77
size dimension, 56
sketching, 14
slot, 6
snap head, 81
socket head, 79
specification, 10
sphere, 7, 59
spigot, 6
split pin, 80
spokes, 6
spot face, 6, 7
spring washer, 84
spur gear, 9
standard part, 10
star washer, 84
stud, 77, 80
sub-assembly drawing, 1
surfaces in space, 48, 50
symmetrical parts, 45

tab washer, 84
taper, 6, 59
taper pin, 80
technical terms, 6
temporary fastening, 78
third-angle projection, 31–2
thread, 8, 77, 79
title block, 2, 3
tolerance, 7, 62–6
tracer, 12
transition fit, 63
two-start thread, 77
types of line, 4

undercut, 7

vanishing point, 15
vertical plane, 26, 31, 48, 50
views on drawing, 45

washer, 40, 84
web, 6, 39, 40
welding symbols, 82–3
wire locking, 84
working limits, 65

xerographic process, 10

zero line, 62